T0338565

Language and Cognition

Essays in Honor of Arthur J. Bronstein

Library of Congress Cataloging in Publication Data

Main entry under title:

Language and cognition.

Includes bibliographical references and index.
1. Language and languages — Addresses, essays, lectures. 2. Linguistics — Addresses,
essays, lectures. 3. Bronstein, Arthur J. I. Bronstein, Arthur J. II. Raphael, Lawrence
J. III. Raphael, Carolyn B. IV. Valdovinos, Miriam R.
P26.B77L36   1984   410   83-22987
ISBN 0-306-41433-3

©1984 Plenum Press, New York
A Division of Plenum Publishing Corporation
233 Spring Street, New York, N.Y. 10013

Printed in the United States of America

Arthur J. Bronstein

Tabula Gratulatoria

Albert Abrams

Belle Adler

Hedda Aufricht

Maria C. Balbi

Asher Bar

Clarence Barnhart

Bernard H. Baumrin

Albert Bermel

Cordelia Brong

Margaret Bryant

Francis A. Cartier

Jean M. Civikly

Carmen De Zulueta

Frances S. Dubner

Wm. Walter Duncan

Abne Eisenberg

Shirley Flint

Marie Fontana

Frances Freeman

Mary and Richard Geehern

Carolyn C. Gilboa

Wilbur Gilman

Alan C. Goodman

Gloria B. Gottsegen

Murray Hausknecht

Beryl F. Herdt

Emita B. Hill

Marnesba D. Hill

Ursula F. Hoffman

Elizabeth B. Holmes

Jacob Kabakoff

Cornelius Koutstaal

Rose Kove

Barbara Kruger

Fred Kruger

Joseph C. Landis

Patricia Launer

Telete Lawrence

Gayle and Mitchell Levison

Johanna Meskill

Gloria Mihàlyi Solomon

Beth Morris

Alan Nichols

Glenn T. Nygreen

Lucille Okoshkia

Georgiana H. Olwell

Mary Pettas

Irwin Polishook

Paul Proskauer

David and Melissa Raphael

Allen Walker Read

Chester H. Robinson

Beulah F. Rohrlich

Olga Romero

Irwin Ronson

Davida Rosenblum

Herbert Rubin

Marilyn Silver

Robert Sonkin

Joel Stark

Jess Stein

Norma K. Stegmaier

Bea Stocker

Blanche R. Teitelbaum

Juergen Tonndorf

Laurence Urdang

Jack G. Valdovinos

Juan F. Villa

Barry Voroba

Joella Vreeland

Ruth Walker

Hollis White

Murray N. Wortzel

Annette R. Zaner

Ernest Zelnick

Contributors

WALTER S. AVIS
Late of Royal Military College of Canada

EDWARD H. BENDIX
Hunter College and the Graduate School of the City University of New York

FREDERICKA BELL-BERTI
St. Johns University and Haskins Laboratories

MOE BERGMAN
School for Communication Disorders, Sackler School of Medicine, Tel-Aviv University

JOHN W. BLACK
Professor Emeritus, Department of Communication, the Ohio State University

DWIGHT BOLINGER
Professor Emeritus, Department of Romance Languages and Literatures, Harvard University, and Professor Emeritus, Department of Linguistics, Stanford University

MICHAEL F. DORMAN
Department of Speech, Arizona State University

JON EISENSON
Professor Emeritus, Department of Hearing and Speech Science, Stanford University

ERICA C. GARCÍA
Department of Spanish and Portuguese, University of Leiden

LOUIS J. GERSTMAN
*Department of Psychology, College of the City of New York of the City
University of New York*

KONRAD GRIES
*Professor Emeritus, Department of Classical and Oriental Languages,
and the English Language Institute, Queens College of the City University of New York*

KATHERINE S. HARRIS
*The Ph.D. Program in Speech and Hearing Sciences, the Graduate
School of the City University of New York, and Haskins Laboratories*

IRVING HOCHBERG
*The Ph.D. Program in Speech and Hearing Sciences, the Graduate
School of the City University of New York*

EDITH TRAGER JOHNSON
Professor Emerita, Linguistics Program, San José State University

D. TERENCE LANGENDOEN
*Linguistics Program, Brooklyn College, and the Ph.D. Program in Lin-
guistics, the Graduate School of the City University of New York*

SAMUEL R. LEVIN
*Department of English, Hunter College, and the Graduate School of
the City University of New York*

HARRY LEVITT
*The Ph.D. Program in Speech and Hearing Sciences, the Graduate
School of the City University of New York*

SAMUEL LIEBERMAN
*Late of Department of Classical and Oriental Languages, Queens Col-
lege of the City University of New York*

RAVEN I. McDAVID, JR.
*Professor Emeritus, Departments of English and Linguistics, University
of Chicago and Editor-in-Chief, Linguistic Atlas of the Middle and
South Atlantic States*

DONALD MELTZER
London and Oxford, The United Kingdom

PAULA MENYUK
Division of Reading and Language Development, Boston University

JOHN B. NEWMAN
*Professor Emeritus, Department of Communication Arts and Sciences,
    Queens College of the City University of New York*

MARDEL OGILVIE
*Professor Emerita, Department of Speech and Theater, Herbert H.
    Lehman College of the City University of New York*

LAWRENCE J. RAPHAEL
*Department of Speech and Theater, Herbert H. Lehman College of the
    City University of New York, and Haskins Laboratories*

MARCIA RICH-SIEBZEHNER
*Department of Psychology, College of the City of New York of the City
    University of New York*

NORMA S. REES
*Vice Chancellor for Academic Affairs, University of Wisconsin-
    Milwaukee*

JOHN F. WILSON
*Department of Speech and Theater, Herbert H. Lehman College of the
    City University of New York*

Preface

We are pleased to be able to honor Arthur J. Bronstein with this volume of essays. We are all the more pleased because the volume has considerable intrinsic merit, but neither the reader nor Arthur should have any doubts about our primary purpose in assembling this book. That the collection is intrinsically valuable is, in itself, a tribute to the man whom it honors: The contributing authors are all colleagues, students, and friends of Arthur.

Readers who are acquainted with Arthur will not be surprised by the broad range of academic expertise which has been brought to bear on the subject of language in this book. They will recognize that Arthur's own range of expertise and interest is only barely matched by the contents of the essays and the backgrounds of their authors.

On the other hand, those who know little about Arthur may have thought of him primarily in narrow association with phonetics and linguistics, most likely as the author of *The Pronunciation of American English,* surely the most influential of American phonetics texts during the last quarter of a century. Although such an association is in many respects appropriate, it is altogether too limited, but this will not deter us from using it as the basis for a relevant and, we hope, revealing metaphor about Arthur J. Bronstein: It occurs to us that Arthur may be thought of in much the same terms as the phonemes and morphemes about which he has enlightened us over the years. There can be no gainsaying that he is a distinctive, significant entity in the academic world. Moreover, it is not difficult to recognize several variants of this entity (dare we say 'allo-Bronsteins'?), each occurring in a well-defined environment.

For instance, there is the administrator-variant, found in the environment of the office: deans' offices at Herbert H. Lehman College and chairmen's offices (Speech and Hearing Science, Linguistics) at the Graduate School of the City University of New York. This variant has always been remarkable for its innovativeness, even-handedness, and ability to motivate colleagues and students to commit acts of scholarship and self-less institutional labor previously thought impossible.

Another major variant is that of the scholar. This is also found in the environment of the office (albeit slightly less spacious than those mentioned above), as well as that of the library, both local and in such exotic climes as Israel, Norway, Peru, and England. No less innovative than its administrative counterpart, the scholar-variant has been employed successfully in work on dictionaries, dialect studies, and textbooks. It is unique in having a strong historical component: It was first observed as an eighteenth-century English vowel but somehow has never been out-of-date, not even when engaged in describing the phonetic exploits of Dionysius Thrax.

Finally, we come to the teacher-variant, found, of course, principally in the environment of the classroom with students at all levels and marked as always by innovativeness, inspiration, intellectual challenge, and challenging standards. Its distribution remains problematical because it is not complementary with the other variants. It is ubiquitous, and in order to achieve descriptive adequacy we must here introduce a transformational component, namely, that any environment ⇒ classroom. Thus, when we observe Arthur J. Bronstein performing as a dean or a chairman, we learn how to administer; when we see him functioning as a scholar, we learn how to go about our research and writing; when he instructs us, we learn how to teach. In short, Arthur continually teaches, sometimes directly, always by example, and, if we pay attention, we learn what we are better off for knowing. For this we offer our gratitude, our respect, and, of course, this book.

LAWRENCE J. RAPHAEL

CAROLYN B. RAPHAEL

MIRIAM R. VALDOVINOS

Acknowledgments

In the course of compiling this book, the editors have been helped in many ways by many people. We want them to know that their assistance has been appreciated. We wish first to thank the authors of the essays, who unselfishly contributed their time and expertise. Those who have experienced that peculiar mixture of joy, labor, agony, frustration, doubt, and satisfaction which authorship comprises will understand the nature of their efforts. To the sponsors of this volume, listed in the *Tabula Gratulatoria*, we owe a special debt of gratitude. Their support enabled us to meet the financial obligations incurred in launching the project and seeing it through to completion.

We were also helped in this regard by the administration of Herbert H. Lehman College and the Lehman College Association. President Leonard Lief, Dean Jack Wiener, Associate Dean Edward Pakel, and Mr. George Metcalf offered us facilities and advice which smoothed our path, particularly in the early stages. We are also grateful to the Lehman College Auxiliary Accounting staff who relieved us of the burden of many record-keeping chores. We have also, thanks to President Kurt Schmeller, made use of facilities at Queensborough Community College, CUNY, where we received encouragement and welcome advice from Professor Sheena Gillespie, Chairperson of the Department of English. In seeking advice about publication, we received valuable assistance from Mr. Jess Stein of Random House and from Professor John B. Newman, Emeritus, Queens College, CUNY. In the final stages of our work we found ourselves indebted to Professor Robert Rieber of the John Jay College of Criminal Justice, CUNY, the general editor of the series of books of which this is one, and to Frank Columbus and Eliot Werner of Plenum Press. Finally, we wish to recognize the understanding and patience of the members of our families, to whom, from time to time, we were less available than we might have wished: Professor Jack Valdovinos, Melissa Raphael, and David Raphael.

THE EDITORS

Contents

Notes on Borrow(ing) Pit

Walter S. Avis
Late of Royal Military College of Canada

I first heard the term *borrow pit* during a visit to Spalding, Saskatchewan in the late 1940s, when the adjacent highway was being rebuilt to modern specifications, asphalt surfacing and all. A short time earlier, my brother-in-law had given the Department of Highways permission to remove from one of his fields near the road a large amount of fill to be used in building up the roadbed. He was left with an enormous pit, which later became valuable to him as a man-made watering-hole for his cattle and a pond for his geese; that is, it became a dugout, of which there are many on the prairies. I later learned that there were thousands of such pits strung along the railways and highways (and nowadays pipelines) of Canada.[1]

Oddly enough, I first saw the term in print a few years later in *Saskatchewan Harvest* by Carlyle King, a revered professor of English at the University of Saskatchewan:

> Overnight the sky traded its winter tang for softness; the snow already
> honeycombed with the growing heat of a closer sun, melted—first from the
> steaming fallow fields, then the stubble stretches, shrinking finally to
> uneven patches of white lingering in the barrow pits. (1955, p. 16)

This passage interested me for several reasons: first, the variant *barrow pit* was new to me at the time; second, it occurred in a nontechnical descriptive context; and third, it was being used freely by a literary man in such a way that he expected it to be understood by his readers. Although I recognized the term as a variant of *borrow pit*, there must have been many other urban Easterners who had no idea of its meaning. Having begun my dictionary work by that time, I made a citation slip for my files. Eventually, both *borrow pit* and *barrow pit* found their way into *The Senior Dictionary* of the Dictionary of Canadian English series (Avis, 1967a). Unfortunately, *borrow pit* also found its way into *A Dictionary of Canadianisms on Historical Principles (DCHP)* (Avis, 1967b), documented weakly by a 1961 citation from the *Canadian Geographical Journal*. Both terms had been included at an early stage because I was ignorant of their provenience and because there were Canadian citations in the files; *borrow pit,* alas, remained in the final draft through an oversight, for I had by then ample evidence to indicate that the term was no Canadianism.

During the later stages of editing the DCHP, I had checked the usual American and British dictionaries: the *DAE, DA, NID, OED,* and *EDD.* Neither *borrow pit* nor *barrow pit* was entered in the first two. The *OED* had *borrow pit* only, in the *Supplement* of 1932, two citations being given, dated 1898 and 1907; the first was from Kipling's *Day's Work,* in a piece relating to India and written in 1893, an antedating picked up in the *Supplement* of 1972, which otherwise adds nothing. Webster's *NID3,* following *NID2,* had *borrow pit* and *barrow pit,* the latter labeled as 'chiefly West'—but, of course, had no information as to the date of the evidence. I also queried my colleagues on the Editorial Board of the *DCHP;* two of these, Douglas Leechman and Charles Crate, had been familiar with both variants in Canada for many years. These inquiries and others indicated that the terms were familiar to civil engineers, who favored *borrow pit* (to them a technical term) and to nonurban Westerners, who knew both variants, some using one and some the other. This unsettled usage was made evident in a conversation I had in 1963 with two middle-aged men from Vanguard, Saskatchewan, lifelong friends and, indeed, traveling together when they visited me in Calgary. One had always used *borrow pit,* the other *barrow pit,* and each was surprised that the other used a different term.

Furthermore, each of these prairie farmers had an explanation for his usage, the first saying that the fill removed from the pit was 'borrowed,' the second that it was, in earlier days, 'wheeled from the pit in barrows.'[2] I might add that the man who said the earth was borrowed was puzzled by the term because there was "no intention of paying it back." Clearly, neither had any idea of the term's origin, and, equally clearly, at least one of the explanations was a popular etymology. Similiar explanations, I might add, were offered by others I queried about this term. Nevertheless, such dictionaries as include *borrow pit* and make reference to its etymology associate it with *borrow* from OE *borige*, for example, *NID3*, which also has *barrow pit* cross-referred to *borrow pit*, identifying the first as a variant of the second. Yet a complication arises at this point, for *borrow pit* is defined as 'an excavated area where material (as earth) has been borrowed to be used as fill at another location,' yet *barrow pit*, though cross-referred to *borrow pit*, is said to mean 'esp: a ditch dug along a roadway to furnish fill and provide drainage.' Also entered in *NID3* are *bar pit*, said to be an alteration of *barrow pit* (and by implication to have the meanings ascribed to *barrow pit*) and *bar ditch*, said to be an alternation of *borrow* (and presumably to have the meaning of *borrow pit* but not those of *barrow pit* and *bar pit*, both of which refer in one sense to a ditch).[3] There is also an entry in *NID3* for *borrow ditch*, the definition being that given above for *barrow pit*.

Wright's *EDD* has an entry for *barrow-ditch*, 'a small ditch.' The term is marked obsolete, located in Kent, and supported by a citation dated 1752 and another, for the variant *barrow dicking*, dated 1784. This particular ditch runs at right angles to a road and a parallel footpath; it serves to drain the road and to keep (to bar?) vehicles from encroaching on the footpath. Wright also has *barrow-mouth* 'an adit or level dug in a hillside.' The *OED Supplement*, I should add, has *borrow-hole:* 1901 *Practioner* Mar. 258, 'Borrow-holes in railway embankments.' This term is entered under and identified with *borrow-pit:* 'In civil engineering, an excavation formed by the removal of material to be used in filling or embanking.' If a borrow-hole is a borrow-pit, then it would appear that here the borrower (burrower?) is robbing Peter to pay Paul, that is, borrowing from an embankment to make another embankment.

However perplexing these gleanings from dictionaries, it is clear that *borrow pit* is the established term nowadays and that *borrow* has also become established, especially in civil engineering, both as verb and noun, which respectively appear thus in *NID3:* 'to dig (as from a borrow pit)' and 'material (as earth or gravel) taken from one location (as a borrow pit) to be used for fill at another location.' The *OED*, by the way, has neither noun nor verb entered in these senses. That the terms are established in civil engineering is evident from various handbooks (e.g., Merritt, 1968). *Barrow pit*, then, must be classed as a regional—and probably obsolescent—term. How, then, did *barrow pit* gain currency among certain speakers, as in the Canadian West—assuming that the antecedent form was *borrow pit*?

One hypothesis might be that a diaphone transfer took place about two generations ago when the West was being settled and the Canadian transcontinental railways being built. For there and then English-speakers of varied dialectal background (Canadian, American, and British) were thrown together. Such a hypothesis would require, say, three groups having different actualizations of the stressed vowels in *borrow* and *barrow:* Group A, having [æ] in *barrow,* [ɔ] in *borrow;*[4] Group B, having [ɑ] in *barrow,* [ɔ] in *borrow;* and Group C, having [æ] in *barrow* and [ɑ] in *borrow.* If it is assumed that *borrow (pit)* was the antecedent form, a speaker of Group A hearing the word pronounced [ˈbɑro] might associate it with his [ˈbæro] rather than with [ˈbɔro], especially in a situation where wheelbarrows were being used to bring the earth from the pit.[5] Once [ˈbæroˌpɪt] had come to be used by a number of people, it would be adopted by still others and passed on to the next generation, becoming an entrenched variant of [ˈbɔroˌpɪt] and developing its own popular etymology. Needless to say, a different process could have taken place if *barrow pit* had been the antecedent form, a possibility that appears unlikely, this form being rare in printed evidence. My earliest and only citation, referred to above, is dated 1955, although the form is certainly much older and more widespread in oral use in the Prairie Provinces at least. Moreover, *barrow* does not occur in either verb or noun senses that are synonymous with those much in evidence for *borrow.*

Having dispensed, after a fashion, with *barrow pit,* I would like to propose that *borrow pit,* now so secure, was not the antecedent form after all. The dated *OED* evidence is scant, 1893 and 1907: *Supplement, 1972;* I have a number of Canadian citations antedating the *OED* and stretching right up to 1977. These were not easy to find, for the term refers to activity of little significance to those who write popular books about railway-builders and last spikes. Borrow pits are of concern to engineers, contractors, foremen, and the navvies who shift the spoil; they are of no great significance to historians, journalists, commissioners, and the tycoons who reap the spoils. Yet it is the last group that has recorded in one way or another the story of the environment where borrow pits are used. To them, the general terms *excavation* or *earthwork* are sufficient to embrace borrow pits, side cuttings, side ditches, side borrows, and ballast pits.

My big problem, however, has not been that of antedating the *OED* evidence but of postdating it. Confining myself to Canadian sources for the time being, I have not been able to postdate the 1893 citation for *borrow pit;* but I did find an instance of *borrow ground* from 1872: "Here some side hill cut and fill of rock is required, 500 cub. yards of rock in all, the rest from borrow ground" in a report by William Murdoch, an engineer working on the Canadian Pacific Railway (Fleming, 1880, p. 303). Interesting as Murdoch's term may be as a near synonym for *borrow pit,* the author of the main report that constitutes the book, Sanford Fleming (later Sir), uses another that proved to be much more significant, namely, *borrowing pit:* "The side ditching and off-take drains must also be grubbed, but no grubbing will be paid for in borrowing pits." (Fleming, 1880, p. 336). This term is clearly a closer synonym of *borrowing pit* than is *borrow ground,* used by Murdoch four years earlier. I decided to follow Fleming's writings backward in time, for he was the most famous and knowledgeable of Canadian railway engineers; originally a Scot, he came to Canada in 1845 at 18, studied civil engineering, eventually being taken on as an engineer (1857–62) by the Ontario, Simcoe, and Hanover (later the Northern) Railway. In 1863 he was engaged to do the first survey for the Intercolonial Railway (ICR), subsequently becoming chief engineer for that line. Established as the prominent railway engineer in Canada, he was later to take over as engineer-in-chief of the Canadian Pacific Railway.

In pursuing further evidence, I next turned to Fleming's earlier historical sketch of the ICR (1876), where I found what I was looking for—and more—on page 119: "the narrow cuttings being insufficient, borrowing pits had to be resorted to," and on page 152, "The quality of embankment required was much in excess of the quality of cutting on the line, and, therefore, extensive borrowing pits were necessary." In the next sentence on page 152, he writes, "In some spots, the material available for borrowing was so scanty that many acres of ground were stripped to furnish the quantity required." *Borrowing* here may refer to the material taken from the pits (a noun) or to the act of digging the material (a verbal), both uses being later recorded for *borrow,* as I have already pointed out. The noun sense is clearer in this sentence from page 185: "Additional borrowing was, however, required for the embankment."

At this stage, I have not had the opportunity to peruse two earlier publications by Fleming. But I did come across in the library of the Royal Military College a most helpful two-volume collection of correspondence, surveyors' and engineers' reports, appendixes to public documents, and miscellaneous papers. This collection has no title page, but a handwritten note on a blank page at the front of each volume identifies the persons who put the papers together and had them bound, namely, D. Pottinger, whom I have not yet identified. The title is simply: *I.C.R., H. & W., and Pictou Extension, 1844–1860* (Vol. 1) and *I.C.R., 1861–67, N.S. Railway, W. & A. Railway, 1861–67* (Vol. 2). There is no publisher identified, but internal evidence implies certain Nova Scotia government agencies; there is, moreover, no date of publication, although it must have been after January, 1868, the date of the last document bound therein. For my citation slips, I have arbitrarily ascribed the date 1869 for want of a better. Nevertheless, almost all of the documents are dated or identifiable as to date. Finally, the pagination is that of the various appendixes, reports, and the like, so that the problem of establishing the appropriate page for a given citation was difficult, my solution being to give the number of the Appendix, the title of the document, and so on, as well as the page.

At any rate, these two volumes (Pottinger, n.d.) provided a wealth of evidence for the terms I was seeking. In Volume 2, on page 2, occurs an undated detailed statement of specifications for contractors which includes the term *borrowing pit:* "All excavations . . . in gravel pits, or in borrowing pits and depot grounds . . . shall be executed and the material deposited according to the direction of the Engineer." This document was drawn up by none other than Sanford Fleming, recently arrived in Nova Scotia, who stated in a letter (p. 23) dated April 6, 1865: "On 25 June [1864] I furnished a detailed specification for the clearing, grubbing, grading, masonry, and other works required for the construction of the line." Thus 1864 is the earliest citation I yet have for *borrowing pit,* that is, 29 years earlier than the *OED* citation for *borrow pit.* It is not, however, the earliest occurrence of *borrowing,* which appears often in several senses near the end of Volume 1 in a document dated 1858 and titled "Railway Committee Evidence," a transcription of hearings wherein contractors are claiming for extra costs which they attribute to misjudgments on the part of the railway engineers. Herein is recorded the trade jargon of the contractors and works foremen themselves. The earliest, referring to 1856, is for *borrowing,* that is, 'the earth brought in' (p. vi): " He was to be paid as if he had taken out the whole quantity, less our borrowing, which we were to be paid for at 2s. 3d." *Borrowing* in this sense (a common sense of *borrow* today) occurs throughout this document with reference to the period 1854–58. Another sense is used in evidence given by John Blackie (p. v): "All I ask for is *extra* borrowing"—where the word refers to payment for fill carried but not adequately allowed for. *Borrowing* is used in still another sense by Thomas McKenzie, a foreman, discussing a contract accepted in September, 1854: "The embankment No. 3 was made from borrowings at each end of the line, of stone and earth. . . . I looked after the work." Here, *borrowing* clearly means 'borrowing pit'—another sense in which *borrow* is used today, as defined in *Funk and Wagnalls Standard Dictionary* (1965): 'a place . . . where material is removed to be used as filling elsewhere.' Oddly enough, this dictionary does not have *borrow pit* as an entry.

In the 1858 document being examined here, the substantive used is *borrowing,* which combines later with *pit* to form *borrowing pit.* Also in the evidence is the verb *borrow,* as used by Thomas McKenzie, foreman (p. xx): "We had to borrow owing to the embankment not being finished, so that we could carry it over. We borrowed at the south end"; and by Eliakim Creelman, a contractor (p. xx), referring to work done in 1855: "We had to borrow about 7000 yards in embankments, about Barrack hill.

If the survey was correct, we would not have had any borrowing, the cutting would have been sufficient." The verb *borrow* remains current in a similar sense, 'take from elsewhere' (as in 'to borrow words') used not only with reference to plain fill but also for ballast, as indicated in this citation from the evidence (p. xv): "The ballasting for No. 5, was not taken out of that cutting—it was borrowed chiefly." It seems probable that evidence exists for *borrowing* and *borrow* earlier than in the citations I have offered from Canadian sources, for railways were being built in the United Kingdom and the United States before construction in Canada got underway.

In the meantime, on the basis of evidence presented here, I would like to offer the following hypothesis as to the etymology of the terms under discussion. The common denominator being accepted as the verb *borrow*, a verbal substantive *borrowing* developed in the context of railway construction, subsequently appearing as an adjunct in *borrowing pit*. Later in the nineteenth century, *borrow* made its appearance as the first component of the compound, apparently displacing *borrowing* in combination with *pit*. This development is common enough, as when 'clothes for working in' gave *working clothes*, which was later competing with *work clothes*. A striking example of this process may be seen in the OED evidence for *bathing tub*, the earliest citation for which is dated 1583, as opposed to *bathtub*, dated 1887. A similar instance relates to *dancing hall* (*OED*, 1753) and *dance-hall* (*OED Supp.*, 1858), the latter being labelled 'U.S.' That compounds of the type $-ing + N$ occur much earlier in the history of English has been pointed out by Marchand (1969, pp. 69–70); *borrowing pit* belongs to his type *writing table:* 'B denoting a place designed for the performance of the action underlying A.' *Borrow pit*, on the other hand, belongs to Marchand's type *bakehouse*, 'vb stem determing sb'—a much later although parallel development in the compounding process and one having a similar denotation (1969, pp. 72–73). The advent of *borrow* as a noun appears to be a relatively recent backformation from *borrow pit*, postdating the use of *borrowing* in a similar nominal sense.

Although my Canadian evidence for *borrowing* (*pit*) is unique and much earlier than that for *borrow* (*pit*), I suspect that the earlier term was born in the railway-construction business in the United Kingdom, perhaps being taken over from mining jargon. It was then brought to North America by British engineers and contractors, who were prominent in railway building on this continent. After all, many railway terms originated in the United Kingdom, as railways themselves did. Many of them were borrowed, often as translation loans (calques), into other European languages. For example, the French equivalent of the term I am discussing is *chambre d'emprunt*, which could translate *borrowing pit*, as is quite likely (the date considered) in the 1874 citation relating to railway construction given in Littré (1956).[6] Modern translation dictionaries of French and English gloss *chambre d'emprunte* as 'borrow pit,' as might be expected. Moreover, the Italian equivalent is *cara de prestito*, another calque; the Spanish is *préstamo* 'a borrowing.'[7]

The foregoing hypothesis is admittedly inconclusive, for a great deal of searching must be done to determine where and when both *borrowing pit* and *borrow pit* came into being. For this reason, I have hedged by referring to this paper as a *note*. Since volunteering this contribution, I have discovered more about these several terms than I had thought possible. In fact, my concluding hypothesis could not have been made two weeks ago, when I was not even aware of the existence of *borrowing* or *borrowing pit*. Perhaps I shall have occasion to take up this complex and frustrating search again; but for the present I must say that I sympathize with those disgruntled contractors working on the Provincial Railway in Nova Scotia a century and a quarter ago; for they also learned that "who goes a-borrowing, goes a-sorrowing."

NOTES

1. See *The Canadian Geographical Journal*, July 1967, 6.3: "The building of highways has both beneficial and adverse effects [on the duck population]. Some ditches and many borrow pits act as artificial potholes."
2. This familiar sight of bygone years is preserved by a photograph which I have seen in several places, the most recent being *The Canadian Geographical Journal*, January, 1961, 15. These stationmen also have their monument: "[A] magnificent cairn [was] erected ... to the memory of the pick-shovel-and-wheelbarrow brigades that built the railway. Its bronze tablets bear the words of Kipling's 'Sons of Martha'—that taut tribute to the sometimes forgotten legion of those who move mountains otherwise than by prayer." (Stevens, 1962, Vol. 2, p. 449).
3. As with *barrow pit, NID3* labels *bar pit* and *bar ditch* "chiefly west." The last-mentioned term is attested by Robert C. Cowser (1963) of Texas Christian University. The second term is said to be "a corrupted form" of the first, which is not identified with *barrow pit;* but then *NID3* does not make the identification either, in spite of the identical definition in both places. Cowser's brief note is of no help in illuminating the complex problem under discussion here.
4. The present generation of young people are moving toward [ɛ] in such words as *marrow, barrow, Harry, marry*. The transfer here proposed took place some two generations ago before the displacing of [æ] by [ɛ] in such words was as widespread as it is now in Canada.
5. This substitution of [æ] for [ɔ] is evident in this citation by Allen (1961, p. 103): "We said 'I barrows first,' for first turn at bat ..." When I was a kid in

Toronto, we said "borrows"; but Allen came from the east end of the city. Similar evidence would not be hard to find. The *EDD*, under *barrow-pence*, has a citation spelled *borrow-pence* ("coins found in a tumulus"), an instance, I assume, of the reverse process.

6. The French calque does not appear in Littré's (1873) *Dictionnaire Française*, having been picked up and supported with a citation by later editors. It does appear in Hatzfeld and Darmestetter's (1964) French dictionary; whether it appears in the 1871 version I have not yet been able to determine. The Spanish *Diccionario de la Lengua Española* gives a definition of *préstamo* equating with the meaning of *borrow(ing) pit:* "un camino . . ." The German *Materialgrube* is outside the calque pattern, judging from the gloss for *borrow pit* in *Maret-Sander Encyclopoedic English–German Dictionary*.

7. The translation dictionaries in which I found these Romance equivalents of *borrow pit* are as follows: Daviault (1945); *The Follett/Zanichelli Italian Dictionary;* and Williams (1956).

REFERENCES

Allen, R.T. *When Toronto was for kids.* Toronto: McClelland and Stewart, 1961.

Avis, W. S., and others. *The senior dictionary.* Toronto: Gage, 1967.(a)

Avis, W. S. (Ed.). *A dictionary of Canadianisms on historical principles.* Toronto: Gage, 1967.(b)

Cowser, R. C. From 'borrow ditch' to 'bar ditch.' *American Speech* 1963, *37*, 157.

Daviault, P. (Ed.). *Military dictionary, English–French, French–English.* Ottawa: King's Printer, 1945.

Diccionario de la lengua española. Madrid, 1956.

Dictionnaire française. Paris, 1873.

Fleming, S. *The intercolonial. A historical sketch of the inception, location, construction and completion of the line of railway uniting the Inland and Atlantic Provinces of the Dominion, with maps and numerous illustrations.* Montreal: Dawson Brothers, 1876.

Fleming, S. *Report and documents in reference to the Canadian Pacific Railway.* Ottawa: Department of Railways and Canals, 1880.

Follett/Zanichelli Italian dictionary. Chicago: Follett, 1968.

Hatzfeld, A., & Darmestetter, A. *Dictionnaire général de la langue française.* Paris, 1964.

King, C. *Saskatchewan harvest.* Toronto: McClelland & Stewart, 1955.

Littré. *Dictionnaire française.* Paris, 1873.

Marchand, H. *The categories and types of present-day English word-formation. A synchronic-diachronic approach* (2nd ed.). Munich: C. H. Beck'sche, 1969.

Maret–Sander encyclopoedic English-German dictionary. New York, 1931.

Merritt, F. S. (Ed.). *Standard handbook for civil engineers.* New York: McGraw-Hill, 1968.

Pottinger, D. [Reports, correspondence, and miscellaneous papers relating to the] *I*[nter-] *C*[olonial] *R*[ailway], *H*[alifax] *& W*[indsor], *and Pictou Extension, 1844–60* (Vol. 1); and *I.C.R., 1861–67, N*[ova] *S*[cotia] *Railway* [and the] *W*[indsor] *& A*[nnapolis] *Railway, 1861–67* (Vol. 2). Halifax: after 1869.

Stevens, C. R. *Canadian national railways.* Toronto: Clark-Irwin, 1962.

Williams, E. B. (Ed.) *The Holt Spanish and English dictionary.* New York: Holt, 1956.

2

The Metaterm 'Cause'
Exploring a Definition in Newari and English

Edward H. Bendix

Hunter College and the Graduate School of the City University of New York

INTRODUCTION

The purpose of this discussion is to explore a particularly flexible but explicit definition for the linguistic term 'cause' or 'causative' as a semantic notion for language-specific description, cross-linguistic comparison, and general theorizing and linguistic dialogue. Examples will be brought in from Newari (Tibeto-Burman) of the general variety used in Kathmandu, Nepal, as well as from English. Strong constraints cannot be built into the definition in the absence of agreed-upon language universals justifying such constraints. Its flexibility lies in allowing an indefinite number of details to be made explicit, and its explicitness, in specifying where those details belong in relation to one another.

The considerable amount of interest shown in a causative of cross-linguistic applicability since the mid-1960s has led to its share of disputes. In these it has too often been unclear whether people were in fact in disagreement or only thought they were when they used 'causative' in trying to communicate. It has often been unclear whether they disagreed about the proper definition of this metaterm, about data and their interpretation, about the proper theory to account for the data, or about the proper

11

treatment of the causative within the framework and toward the goals of a particular syntactic theory.

Part of the difficulty in knowing what we mean by the causative lies in its history. The term has roots in the description of languages with one or more morphemes called 'causative,' typically expressed as verbal affixes. In its linguistic origin, then, it is primarily a morphological label. Semantically, it has not been specific enough to be more than suggestive of a meaning, whether with regard to a particular language or for more general linguistic discussion. Often it means, to a given linguist, what the morpheme so labeled seems to mean in the language or languages studied by that linguist, so that in scholarly dialogue languages may be lumped together that actually differ as to whether their respective causative morphemes more broadly cover general transitivity, denominative verb derivation, both 'to enable' as well as 'to cause', and so forth, or more narrowly exclude some of these. What is needed, then, is a definition for our linguistic metalanguage that allows us to be explicit in our identification of variables and of potential universals.

Since the issue in this exploration is semantic, it will not matter whether in a given language the causative is associated with a morpheme or a construction, nor, if with a morpheme, whether that morpheme is a verbal affix, a separate verb, or a subordinating conjunction, nor whether a given morpheme can be described as having causative as really only one of its functions. Such concerns require separate treatment.

Given the concentration on a semantic matter, this essay also will not be the much longer discussion it could be if it were to give extensive consideration to the body of literature that has been produced on the subject of the causative. Most of this literature has syntactic or combined semantic and syntactic concerns in which workers have frequently differed on what to attribute to semantics and what to syntax, when the distinction has been kept. A good example of the range of approaches and interests may be found in the collection of articles edited by Shibatani (1976b). These articles, not to mention their bibliographical references, can be consulted for their insights revealing many more of the details, fine distinctions, and variables that, when they have to be specified, could be accommodated by the flexibility of our definition.

Reference to that collection of articles is not meant to endorse everything found there. Without going into detail, we can still say that the collection reveals our present lack of any agreed-upon theory for distinguishing communicated meaning from pragmatic inferences. Particularly unfortunate, therefore, is the frequent merging, if not confusion, in some articles, of the question of a meaning for the causative with beliefs and facts about the world, with details of particular hypothetical situations, and with causality as a philosophical notion. We must maintain a conceptual distinction between what we claim is actually communicated by linguistic forms and the endless detail that can be supplied by the users of those forms in the process of interpretation. Since reference is a pragmatic activity performed by the users, our definitions of causativity do not make reference to causation. Only users can try to make a match between a particular token of a causative expression and a particular relation between states of affairs that can be considered a causal one. Only users can disagree about whether a given token acceptably characterizes a given situation. We can change our minds about the nature of causality without changing our linguistic meanings.

THE DEFINITION

One recurrent conclusion that has consistently appeared most useful is that the causative should be seen as allowing reference to a relation between events and/or states of affairs (see also Bendix, 1966, pp. 62–63). This has been called the 'bisentential' view (e.g., Wojcik, 1976, pp. 170ff), among other labels. Thus, the formulation of the metaterm will be as 'cause' linking two propositions 'P' and 'Q'. The notational devices are not so important so long as we know what is being talked about: P and Q could be labeled 'S_1 and S_2';'P causes Q' could be formulated with predicate or verb first as 'CAUSE $[P, Q]$'. For clarity and consistency, P will always be used for the causing event or state and Q for the caused one.

The flexibility of the formulation resides in the fact that one may specify the propositions P and Q in as little or as great detail as descriptively necessary. Adapted from logical notation, P or Q is a proposition which is expanded to an argument and a predicate. An argument and its predicate will be roughly expressed in English as a subject noun phrase and its predicate verb phrase but are not identical with them. When it is a relation, a predicate may have a second argument. Such a predicate and second argument can be expressed as a transitive verb plus its object noun phrase or similar relational expression. P and Q need not be specified as to whether a predicate represents a state or an action where data do not require it. ('Action' includes such aspects as inchoative and process). If a predicate represents an action, an argument of the predicate may be specified as either performing or undergoing the action. An argument of P and an argument of Q may be specified as identical, such as in a reflexive. P and Q can be complex propositions so that the nature of the relation between an argument of P and one or another argument of Q can be given. For example, the arguments can be specified as being referable to entities that make direct contact (Saksena, 1982). The relation between the time reference variable of P and the time reference variable of Q may be stated or not.

Nonspecification, such as not saying whether a predicate is a state or an action, should be taken to represent vagueness or generality as opposed to ambiguity or polysemy. True ambiguity or polysemy should be represented by disjunctive specification, such as definition of a predicate as either a state or an action. In any case, there should be good reasons for concluding that there is ambiguity. It should arise from the data of the language and not, as frequently happens, from the metalanguage of description, from beliefs about the world, or from other sources more properly handled by the pragmatics of inference. Of course, we might ultimately get evidence for the claim that certain categories are universally distinct and always give rise to ambiguity where a language does not mark the distinction. Such might indeed be the case for state as opposed to action, but even then it is not clear whether such a universal human concern should not be treated as falling under a universal pragmatics of interpretation rather than under semantics, so that the ambiguity is not a semantic one. Without principles for identifying it in the data, ambiguity or polysemy could be multiplied indefinitely, subject only to the limits of imagination (Weinreich, 1980, pp. 118–119).

THE DEFINITION APPLIED TO SOME NEWARI AND ENGLISH CASES

Most of the rest of this discussion will apply the 'P causes Q' formulation to some illustrative data collected from speakers of Newari and, to a lesser extent, to some English examples.

A Question of Scope in Newari

Since the first case involves the scope of Newari evidentials over P and Q, they will require a brief description. Newari has obligatory evidential categories, two of which will concern us here. These two are expressed by suffixes to the verb which also express tense-aspect. With one suffix, the speaker communicates that the evidence for assertion of the event reported is external to the speaker and incontrovertible (as opposed to hearsay, for example) and most commonly involves direct observation of the event. This is the -o suffix in (1). With the other, the evidence is internal, involving conscious, intentional performance of an action, the -a in (2).

    (1) wo-ń yat-o[1]
        he-agt do-ext
        He did (it).

    (2) ji-ń yan-a
        I-agt do-int
        I did (it).

    (3) wo-ń yan-a dhal-o
        he-agt do-int say-ext
        (He) said he did (it). (he = he)

    (4) wo-ń yat-o dhal-o
        he-agt do-ext say-ext
        (He) said he did (it). (he $\neq$ he)

Such internal evidence can only be experienced by the performer of an action, and only the performer uses the internal evidential suffix on the verb in reporting the action, as 'I' in (2) or 'he' in the indirect discourse of (3). In (4), where 'he' $\neq$ 'he', the report of another's action requires -o on the verb expressing the action.

    In the case of causatives, these two evidentials are interpretable with varying scope over the propositions P and Q. For example, the verb *kur-k(ol)-*, translatable as 'to drop, knock or throw down', may be glossed for morphological and other reasons as 'cause to fall'. In the assertion of an action, the usual evidential for the first person is the internal if the action is not defined as involving lack of conscious control on the part of the agent. Thus, (5)

    (5) ji-ń wo kur-k-a[2]
        I-agt it down-caus-int
        I knocked it down.

yields the translation 'I knocked it down' and the information, supplied by the internal suffix -a, that this action was intentional on my part. Since, with the causative suffix -*k(ol)*-, P is always an action, we can analyze (5) as: *P*-'I did something' causes *Q*-'it fell down', What is not specified by -a is whether the scope of my intention is the whole action of knocking down or only the action *P*-'I did something', that is, whether or not my intention included *Q*, that the thing fall down. The dictates of one's theory will determine whether this indeterminacy of scope is an ambiguity and where and how such ambiguity is to be accounted for. As mentioned above, pragmatic rather than semantic ambiguity appears to be the most productive assumption. That scope of intention is indeterminate without further help from the context is shown by (6). Here, the internal, intentional evidential -*a* is

(6) ji-ń mo-si-sse kur-k-a
    I-agt not-know-ing down-caus-int
    I knocked (it) down accidentally.

accompanied by an adverbial translatable as 'unknowingly, accidentially'. Although there is an apparent contradiction between 'intentional' and 'unknowing, accidental', one easy speaker interpretation is that I accept responsibility or blame myself for the action, derived from the internal suffix, but apologize, derived from the adverbial. This interpretation can be led back to an analysis that the scope of the internal evidential is only P–'I did something', whereas the scope of the adverbial 'accidentially' is the whole relation 'P causes Q'—my intentional action had an unintended result.

The internal can probably be said always to cover the causing action P necessarily, as the action I personally perform. Any extension of the scope of this internal intention to the caused event or state Q we attribute to inference. However, the same cannot be said for the external evidential. 'I knocked it down' can also be expressed with the external suffix -*o* as well as with the internal -*a*, as in (7).

(7) ji-ń kur-kol-o
    I-agt down-caus-ext
    I knocked (it) down.

Here I indicate external, observational evidence for the report of an action which I myself performed. The selection of -*o* where -*a* is both possible and the more expected suffix has various uses. One speaker interpretation of (7) is that only later when the action was complete did I notice what I had done. The scope of the external evidential apparently is difficult to narrow down only to the proposition P since adding 'unknowingly, accidentally' to (7) as in (8) now yields a sentence judged to be strange.

(8) (?) ji-ń mo-si-sse kur-kol-o
       I-agt not-know-ing down-caus-ext
       I knocked (it) down accidentially.

One reason given is that (7), that is with the external evidential suffix alone, already implies 'accidentially'. The strangeness might be rendered by the English sentence '(?) I unintentionally knocked it down accidentally'. Once I indicate external evidence for the report of my own action and this is interpretable as referring to lack of intention or accidentalness—a reasonable inference with such actions as knocking things down—a further specification within the same clause that this accidental action was done unknowingly or accidentally approaches incoherence. This further specification suggests that I am adding new information, namely that among actions performed unintentionally or accidentally I am further specifying one done unknowingly, which is uninterpretable without starting the interpretation process all over again and finding a pragmatic context in which to embed the sentence and which will make sense of it. If I merely wanted to emphasize the unknowingness or accidentalness by reasserting it rather than by making it new information, a different syntactic formulation would be needed.

For (6), the apparent contradiction between the intention conveyed by the internal evidential and the accidentalness expressed by the adverbial was resolved by limiting the scope of the internal to P and letting 'accidental' cover the whole relation 'P causes Q'. It appears that the same cannot be done for (8), namely, to predicate 'unintentional/unknowing/accidental' first of P and then as added new information to predicate it also of the whole event. Without dwelling on an account of this in step-by-step argument, one can say that if the causing action was performed by accident, then the result cannot have been intended to be produced by that action but is an accident as well. Alternative accounts are possible in terms of the relation between observed evidence (was only Q observed, or only part of Q?) and the assertion of the complete action 'P causes Q' on the basis of such evidence.

The Newari Causative

So far we have explored the usefulness of the 'P causes Q' formulation in allowing us to talk of the scope of evidentials and of the link between the question of scope and the identity of the asserter with the subject argument or agent of P in the case of the internal. Next we will look at the Newari causative itself, including the question of an identity relation between the agent of P and an argument of Q.

The causative morpheme -k(ol)- is suffixed to the verb stem. In this way, the transitive verb hiõ-k(ol)- 'cause to turn or change' is derived from the intransitive hil- 'turn or change'. The transitive ciõdhon-k(ol)- 'make small' from the intransitive ciõdhon- 'be/become small', lhwon- k(ol)- 'make stout or fat' from lhwon- 'be/become stout or fat', and so forth. The causative morpheme can be added to simple transitive verbs as well. Another feature of the language is that noun phrases, including pronouns, need not be overtly expressed in sentences, their absence generally being interpreted as anaphora for verbs requiring an agent, patient, and/or object role. This gives rise to a problem with causative verbs derived from intransitives when the 'object' (or patient) noun phrase is not expressed. Such causative verbs are of two kinds. The first, to be called 'transitive' causatives here, is exemplified in (9). Sentence (10) is an example of the second kind, to be called 'neutral causatives for the purposes of this discussion.

    (9) ji-ń civ́dhon-k-a
        I-agt small-caus-int
        I made (it, him, her, etc.) small or humble.

    (10) ji-ń lhwon-k-a
         I-agt fat-caus-int
         I made (. . .) fat.

In (9), with 'object' noun phrase not expressed but 'understood', the interpretation is that I act on something or someone other than myself. However, (10) may be interpreted either as my being involved in becoming stout or fattening myself, without an understood object noun phrase, that is as an inherent reflexive, or as my making someone else stout, in which case an object noun phrase is understood. The same would hold for (9) and (10) with other persons as agent. Using contextual cues or expressing the object noun phrase can, of course, clarify the intended interpretation. In addition, for such neutral causatives, an explicit reflexive pronoun as object noun phrase is possible. Added to (10) for example, an overt reflexive may be interpreted as emphasizing my acting upon myself, such as with regard to the quality and quantity of food I put into myself, as opposed to the inherent reflexive reading. In sum, when causative verbs derived from intransitives occur without an overt object noun phrase, the transitive causative verbs are interpreted nonreflexively, whereas the neutral causatives may be interpreted either reflexively or nonreflexively. The transitive causatives require an explicit reflexive pronoun to express reflexivity.

    Causative affixation, being a derivational process, offers no guarantee that the meaning of a derived verb is predictable by linguistic rule from the meanings of its parts. In terms of the 'P causes Q' formulation, the proposition P of derived causatives is 'x does something'—that is, always an action in Newari. Q is, for example, 'y becomes small' or, in a different verb-first notation, 'BECOME [SMALL [y]].' Whether the inchoative 'become' must be made explicit here will be taken up below. We must now specify further the relation between y and the agent x in order to make the distinction between the two kinds of causative verbs.

Several alternative accounts present themselves, and a broader theory would have to decide between them. One account would say that the question of identity between x and y is not specified for the neutral causatives, which do not need an explicit reflexive pronoun to be interpreted reflexively. That is, either '$x = y$' or '$x \neq y$' may be interpreted in the absence of an explicit object noun phrase. The transitive causatives are specified as '$x \neq y$', which takes care of their consistently being interpreted as not reflexive without an object noun phrase. The reflexive pronoun explicitly needed by transitive causatives to express reflexivity is defined as specifying that x and y are different aspects of the same entity, aspect x as agent acting on aspect y. An example is (11) with explicit reflexive, translatable as 'Ram made himself small (i.e., appear humble)'.

(11) ramo-ń thov́-to civ́dhon-kol-o
      Ram-agt self-dat/acc small-caus-ext
      Ram made himself small.

Here, one part of Ram, that having control (Givón, 1975), acts on another to achieve the effect of appearing small.

An alternative account would say that the neutral causatives are ambiguous rather than just general. Then, instead of leaving the relation between x and y unspecified, one makes the specific disjunctive statement '$x = y$ or $x \neq y$'. Or one can say that each verb of the neutral causatives is really two homonymous verbs. One of the two is a regular transitive causative which, like other transitives, may have its object noun phrase understood. The other of the two is an intransitive causative, perhaps derivable from the transitive, and is inherently reflexive, that is, '$x = y$'.

As a last alternative to be considered, one may say that there is no distinction between the transitive and the neutral causative verbs in the first place. All are unspecified as to whether '$x = y$' or '$x \neq y$'. The distinction lies outside of linguistics in the beliefs about actions held by speakers of the language. If '$x \neq y$' is consistently interpreted for a given verb, what we have called transitive, it is because the class of actions to which the verb is used to refer, and not the verb itself, is seen as one in which agents ordinarily act on someone or something other than themselves, or, if in the explicit reflexive, on a part of themselves other than the controlling agent part. When either '$x = y$' or '$x \neq y$' may be freely interpreted, it is because the actions referred to may be performed either on oneself or on others. Of the several alternatives, the last is the one most accessible to empirical testing.

'Make' and/or 'Let' in the Causative

Next we briefly consider the element 'cause' itself. As in some other languages, Newari derived causatives typically, although not always, cover both the senses 'to cause or make' and 'to let or enable'. Needless to say, the English words 'cause', 'make', 'let', and 'enable' are used here as convenient translations or glosses, and the element 'cause' in 'P causes Q' is borrowed from English without its English meaning intact to label the metaterm we are trying to define. In this broader sense of 'make/let', (10) above, in its reflexive interpretation, may be translated not only as 'I fattened myself, made myself stout, built myself up' but also as 'I let myself get fat'. Also transitively interpreted, the verb in (10), *lhwon-k(ol)*-, may be 'x made y fat' or 'x let y get fat'. A further example of the broad range of the Newari causative is (12):

(12) mamo-ń moca.ya-to duru twon-kol-o
      mother-agt child-dat/acc milk drink-caus-ext
      The mother gave the child milk to drink (e.g., in a glass), suckled the child, forced the child to drink the milk, treated the child to (a glass of) milk (e.g., paid for it and let the server give the child the milk), etc.

For Newari causatives, when 'P causes Q' is asserted, Q is also asserted to occur even in contexts where the agent x is interpreted as letting or enabling. For Q not to be specifically asserted, a different construction is needed, as in (13).

(13) wo.y-to woe-k-e biy-a
      him-dat/acc come-caus-inf give-int
      (I) let, made him come.

An awkward translation would approximate (13) somewhat more closely: 'I gave him the causing of his coming'. This still does not specify the sense 'let', being rather more of a benefactive, but since Q–'he comes'—is not directly asserted, although still inferable, the interpretation 'let' is arrived at more easily—he could have refused to come. For specifically rendering English 'let', as in 'I let him come', embedded discourse expressing directive causativity is a common device, for example, the equivalent of English 'I told him he could come'. Using the simple, that is, noncausative, form of a verb in the construction found in (13) has rather restricted uses.

It is not clear how to define this more general causative 'make/let' in the metalanguage so as to distinguish it from the more narrow one 'make', as well as from 'let'. What does seem most promising is that 'P causes Q' in fact be made to represent the more general causative, whereas the more narrow ones would require further specification of P or Q. For example, for 'let, enable, allow' without specified assertibility of an action or state in Q, Q could be specified as containing a modal 'can, possible' as the higher predicate. What *is* thus specified as assertable is that the action or state is or becomes possible.[3] An elegant, alternative solution sometimes offered for rendering 'let, enable, allow' introduces negation into 'P causes Q' and into Q and can be paraphrased as 'x does not cause y not to do or be something'. This formulation, however, does not permit the flexibility needed for specifying whether the agent x does something which then allows Q to occur or whether x does not do something to stop Q or whether a given language does not distinguish these two cases.

Not all Newari-derived causatives cover both 'make' and 'let'. Thus the causative of the verb 'die' stands opposed to a separate verb 'kill', as in (14) and (15):

> (14) ji-ń wo.y-to kasi-i si-k-a
>   I-agt him-dat/acc Kasi-loc die-caus-int
>   I let or had him die in Kasi (Benares), i.e., I took him to Kasi to die.

> (15) ji-ń wo.y-to kasi-i syan-a
>   kill-int
>   I killed him in Kasi.

The causative of 'die' requires an adverbial complement, such as location, *kasi-i* 'in Kasi', or manner, *yaŭko* 'peacefully': y is going to die anyway, and x can only cause it to take place in a certain way. For 'P causes Q' in this case, the assertible proposition Q must be formulated to specify that an adverbial is its predicate. The proposition 'y dies' is only an argument of this predicate but is asserted if the whole proposition Q is asserted— that is, strictly speaking, Q and 'y dies' are necessary inferences if the sentence token is in the assertive mode. However, 'y dies' is the whole proposition Q when one characterizes the separate verb *syat-* 'kill' as 'cause to die', namely '[x does something] causes [y dies]'. It may be that for this Newari verb, as also for '*kill*' in English, it is necessary to specify P further as 'x does something to y' or 'x acts on y', that is, that x affects y directly, as opposed to more general causatives, which are silent on this point. As a simple transitive verb, $syat$-'kill' has its derived causative, as in (16):

(16) ji-ń nae.ya-to/nae-nŏ mev́ sya-k-a
        I-agt butcher-dat/butcher-agt buffalo kill-caus-int
        I had the butcher kill the buffalo/I had the buffalo killed by the
        butcher.

Such a transitive as sya-$k(ol)$- 'cause to kill' must thus be defined as a causative embedded in a causative. Morphologically, the causative morpheme -$k(ol)$- can be suffixed only once to a verb stem. Such a stem can also be one of a handful of verbs which still show vestiges of an older Tibeto-Burman causative prefix (Matisoff, 1976) involving aspiration of initial consonants in Newari, for example don- 'rise, get up, awaken,' $thon$- 'raise, lift, get up (trans.), awaken (trans.),' $thon$-$k(ol)$- 'cause someone to raise something or someone, and so forth.'

Some English Cases

Connected with the ergative nature of Newari grammar is the necessity mentioned above of specifying by some general rule that the proposition P is always an *action* performed by x. English causatives do not seem to need this general restriction. In the English sentence (17), our inference for P is of the terrain as being in a state rather than as performing an action.

(17) The flat terrain let us see for miles.

Similarly, we can define the verb *give* as 'cause to have' and then consider sentences (18) and (19):

(18) Those rosebuds give me an idea.

(19) John gave me an idea.

For P here, the rosebuds are in a state, whereas John is most easily interpreted as performing an action. In English, then, we must be careful not to specify whether the predicate of P is a state or an action without good reason.

The same consideration must be given to the verb *remind* (see Postal, 1970; Bolinger, 1971). We can paraphrase *x reminds y of z* as 'x

causes *y* to think of *x*'. That is, in '*P* causes *Q*', *P* says something about *x* and *Q* is '*y* thinks of *z*.' (Whether *Q* should be an inchoative '*y* remembers *z*' instead of '*y* thinks of *z*' is taken up next below.) The verb *remind* could be considered ambiguous. For example, (20)

(20) The secretary reminded the boss of the visitor waiting outside.

can have the interpretation that the secretary did something such as telling the boss that the visitor was waiting or the interpretation that it struck the boss that the secretary looked like the visitor. In the first interpretation, *P* is that the secretary did something, and in the second, *P* is that the secretary was in some state. We could thus account for the ambiguity by specifying for the verb *remind* that *P* is either an action or a state. One could also claim that there is no semantic ambiguity here, that *P* is simply not specified for action versus state, and that the ambiguity is pragmatic (see page 14 above, on nonspecification). Neither approach in itself handles the cross-over phenomena that Postal tries to account for in his *strike-like* analysis. However, alternative semantic accounts of these phenomena would require a separate discussion.

We can now ask whether, in '*P* causes *Q*', *Q* must always be an action or nonstate. More narrowly, if *Q* involves a state, must it always specify the inchoative of that state? For example, in (21)

(21) Henry lowered the music stand.

can we say that Henry did something that caused the stand to *be* lower or must we say the stand *became* lower? If one must always specify inchoativity for states in *Q*, one really need only specify the state and then have one general rule that inchoative be added in such cases. It is not, however, clear that inchoative is always necessarily there. With *x gives y z* as '[*x* does/is something] causes [*y* has *z*]', consider the two sentences (22) and (23):

(22) Samson's hair gave him great strength.

(23) Popeye's spinach gave him great strength.

(The sentences could also be expressed with the explicit causative *made him very strong*.) Those familiar with these two personages know that as long as Samson had his hair, he had his strength continuously—that is, there is no inchoativity interpreted—whereas Popeye had to eat a can of spinach each time he was faced with a crisis in order to get his great strength for that particular occasion—that is, inchoativity is interpreted.

It is clear that inchoativity is not communicated for Q by the verb *give* but supplied pragmatically. Thus we define the verb *give* as 'cause to have,' which is neutral on the question of inchoativity, rather than as 'cause to get'. In other words, we have an indication that we cannot establish *a priori* whether Q contains inchoative for states in a given verb in a given language.

For simplicity, the examples so far have omitted a necessary specification that will now be taken up, namely the time variables of the propositions in 'P causes Q'.

    (24) John caused Henry to die on Tuesday.

    (25) John killed Henry on Tuesday.

In (24), if the time adverbial *on Tuesday* is interpreted as being in direct construction with the verb *die*, then it identifies when Henry died (Q), and John could have done on some previous day whatever it was that he did (P) to cause the death. Thus, for the English verb *cause*, with Q standing for what is embedded after *cause*, the time adverbial binds the time variables of Q only, whereas it is extendable to the time variable of P only by inference or by interpreting it in construction with the verb *cause*. However, in (25) the adverbial gives us the time of P, John's action, as well as of Q, Henry's dying. With *kill* as 'cause to die', then, we must specify the relation between the time variables of P and Q, thereby making explicit the time distinction between the lexicalized causative *kill* and the construction with *cause*.

The time relation between P and Q may have to be separately specified for all lexicalized causative verbs like *kill* that require it, or it may be predictable by a general rule, one that, say, derives from surface constraints on the scope of an adverbial of time over lexicalized causatives (Fodor, 1970) or involves verbs with an inchoative Q (e.g., the 'die' in *kill*). But such matters need separate study (see, e.g., Shibatani, 1976a). The relation of tense-aspect to the time variables of P and Q might be added here. Thus, the tense-aspect of (26)

    (26) You're killing me.

attaches to both P and Q, and we may paraphrase (26) as 'you are doing something (to me) such that I am dying (am becoming not-alive)'.

Since complex noun phrases can appear in subject position, we must be clear about what P is. This can be shown by comparing (27) and (28), from Fodor (1970, with his numbering supplied), with (29):

(27)[20] John caused Bill to die on Sunday by stabbing him on Saturday.

(28)[21] *John killed Bill on Sunday by stabbing him on Saturday.

(29)   John's stabbing Bill on Saturday (finally) killed Bill on Sunday.

The interpretability of (27) and the difficulty of (28), at least as it stands, support the conclusions drawn from (24) and (25). However, the easier interpretability of (29), if we supply the information that Bill lingered on into Sunday before dying of the condition arising from the stabbing, could be taken to contradict these conclusions. If we took (29) as a counterexample because it contains the verb *kill,* we would be overlooking the fact that not *John,* but *John's stabbing Bill on Saturday,* is the subject of (29). Since the subject of a causative sentence (at least in English) is represented only as an argument within P, 'John stabbed Bill on Saturday' is not represented by the whole proposition P but by an argument within it. This argument and its predicate together constitute P, a proposition very loosely paraphrasable as 'John's stabbing Bill on Saturday did something'. For any given token of sentence type (29), an interpreter can supply a linking course of events that fills in this predicate, 'did something', with as much detail and complexity as necessary to make sense of the token for the occasion of its use. An attempt to formulate a sample predicate with this argument for P discursively in, say, English for a hypothetical occasion of use will reveal two things. First, the proposition P makes no such distinctions as topic and comment since it is part of a definition and not part of a pragmatic discourse; yet a discursive translation of P, as part of a total paraphrase of a token of (29), will inevitably introduce such questions of focus as topic and comment, or questions of time reference, and can lead to endless awkwardness. This is only a warning should such a formulation be attempted. More important, the scope of the time variable of P (as would also be the case for Q) must be free to be narrowed down to any part of P for representing any given token of interpretation. Given that the time variables of P and Q are cospecified for *kill,* an interpreter of (29) will narrow down the time variable of P to that crucial part of the whole inferred process which was there on Sunday to kill Bill.

CONCLUSION

Cross-linguistic comparison of 'causative' requires the semantic description of causatives in different languages by means of explicitly defined technical terminology. We have explored whether 'causative' as a semantic term is most profitably defined in a general sense and as a relation between events or states, that is, 'P causes Q', 'CAUSE $[P,Q]$', and so forth, where the various types of causative across languages and within one language are distinguished by different characterizations of P and Q and of their time variables, P and Q being represented as propositions.

As a formal feature, we will say that the data to be accounted for are bilinguals' assent to, or statements of, translation equivalence of utterances in two different languages, that is, in two different systems of symbols. If both are natural languages, there must be further translation into the symbols of the metalanguage. The "bilinguals" who assent to translation equivalence between a natural language and the metalanguage must be specialists intersubjectively trained in the use of such metaterms as 'P causes Q'.

NOTES

1. Since *wo* is a general third person singular, these sentences could all be translated with 'she' instead of 'he.' 'Ext' stands for 'external evidential,' and 'int' for 'internal evidential.' Other abbreviations are the usual ones: agt = agentive (suffix), caus = causative, dat = dative, acc = accusative, loc = locative, inf = infinitive. Although called *agentive* in these ergative constructions, the suffix -*ñ* does not differentiate between agentive and instrumental, agentive being inferred from animateness or other contextual features. -*ñ* and -*ʊ* are cover terms for a variety of morphophonemes. -*ʊ* is realized as lengthening of the preceding vowel, and -*ñ*, word-finally, as lengthening and nazalization of the preceding vowel. *o* covers a broad schwa range. *dh*, *lh* are voiced aspirate (breathy) *d*, *l*.
2. For *wo* to be clearly interpreted as animate, as in 'I made her/him fall down,' it requires the dative/accusative form *wo.y-to*.
3. 'Specified assertibility' refers to the indication that an element, when it occurs in a clause, will be asserted (together with any other elements so specified) if the clause token is marked for assertion. Thus, if only the possibility of the action or state in Q is specified assertable, then assertion of the action or state itself is not made although still inferable as intended.

REFERENCES

Bendix, E. H. *Componential analysis of general vocabulary*. The Hague: Mouton, 1966.
Bolinger, D. L. Semantic overloading: A restudy of the verb 'remind.' *Language*, 1971, *47*, 522–547.
Fodor, J. A. Three reasons for not deriving 'kill' from 'cause to die.' *Linguistic Inquiry*, 1970, *1*, 429–438.

Givón, T. Cause and control: On the semantics of inter-personal manipulation. In J. Kimball (Ed.), *Syntax and semantics* (Vol. 4). New York: Academic Press, 1975.

Matisoff, J. A. Lahu causative constructions: Case hierarchies and the morphology/syntax cycle in Tibeto-Burman perspective. In M. Shibatani (Ed.), *Syntax and semantics* (Vol. 6): *The grammar of causative constructions*. New York: Academic Press, 1976.

Postal, P. M. On the surface verb 'remind.' *Linguistic Inquiry*, 1970, *1*, 37–120.

Saksena, A. Contact in causation. *Language*, 1982, *58*, 820–831.

Shibatani, M. The grammar of causative constructions: A conspectus. In M. Shibantani (Ed.), *Syntax and semantics* (Vol. 6): *The grammar of causative constructions*. New York: Academic Press, 1976.(a)

Shibatani, M.(Ed.). *Syntax and semantics* (Vol. 6): *The grammar of causative constructions*. New York: Academic Press, 1976.(b)

Weinreich, U. Explorations in semantic theory. In U. Weinreich, *On semantics* (W. Labov & B. S. Weinreich, Eds.). Philadelphia: University of Pennsylvania Press, 1980. (Originally published, 1966.)

Wojcik, R. H. Where do instrumental NPs come from? In M. Shibatani (Ed.), *Syntax and semantics* (Vol. 6): *The grammar of causative constructions*. New York: Academic Press, 1976, pp. 165–180.

3

Assessing the Perception of Speech
A Change of Direction

Moe Bergman

School for Communication Disorders, Sackler School of Medicine, Tel-Aviv University

During the years when I was privileged to serve with Arthur Bronstein and other distinguished colleagues in the activation of the doctoral program in speech and hearing sciences at the City University of New York, I saw my area of specialization, audiology, expand in response to the mutual stimulation its practitioners enjoyed from such provocative company. From a narrow interest in the sensitivity of the ear to sound and its ability to discriminate among isolated words of limited test samples, we were alerted to the complex and multifaceted processes and influences which enter into human communication via speech. The following essay presents, in very abbreviated form, some of the questions we have been asking and some of the findings from various studies undertaken by us and by others.

Just how *do* we understand a spoken message? Do we have to hear the phonemes accurately? The words? How much are we helped by the "music" of speech—its melody and rhythm, its stress pattern and its pauses? How much does visual information contribute? What are the other factors which significantly affect our understanding? How and why do we differ from each other in the perception of speech?

It is readily apparent that we often understand messages which are badly mutilated in their physical attributes. That is, they have been stripped of some of their frequency spectra (e.g., over the telephone); portions may have been obliterated by unfavorable room acoustics or by competing noises or by the distracting conversation of others nearby; or perhaps they were spoken by talkers who have poor voices or deviate speech patterns.

It is remarkable how much alteration of the speech signal our listening skills can overcome. And yet, there are times when the acoustic conditions or the transmitting medium is not the principal reasons for our failure to understand. The trouble, then, seems to be either in us, the listeners, or in the structure of the message itself, that is, in its phonology, its syntactic structure, or its meaning.

The successful transfer of a message from sender to receiver clearly requires a kind of linguistic symbiosis between them. The lack of such linguistic mutuality has plagued the technology of machine recognition of speech, giving rise to such amusing but insightful examples as the story related by Carl Sagan in his charmingly informative paperback *The Dragons of Eden*. A United States senator, being taken through a demonstration of the translation skills of the computer, suggested, upon the invitation of his hosts, the sample message "Out of sight, out of mind." After impressive clicks and flashes, the machine printed out the Chinese equivalent, which was promptly reintroduced to the computer for translation back into English, and which then offered up its startling processed version, "Invisible idiot."

It seems, as William James insightfully observed in 1899, that "much of what we think we see or hear is supplied from our memory. We overlook misprints, imagining the right letters though we see the wrong ones." This is supported, in audition, by studies such as those by Warren (1970) and Warren and Warren (1970), who demonstrated that if a phoneme is deleted from a message it is usually not missed by the listener, who will not even know which phoneme has been removed or from what part of the message.

It is suggested that relatively too much effort has been allocated to the recognition of phonemes in the study of speech perception, whereas more contributing features in the interaction of talker and listener have been sadly neglected. We realize this particularly when we undertake studies of how young children gradually mature in their understanding of speech or how the perception of speech declines, under unfavorable listening conditions, as we reach middle age and beyond. We are then confronted with the need to see the process of speech recognition in its full breadth. Success or failure in phoneme discrimination simply fails to tell us very much of what we need to know.

The successful perception of speech requires, in addition to a commonality of expectancies shared by the sender and receiver, certain abilities in the listener for the processing of a diversity of conditions of the message. Those abilities are dependent upon personal factors such as the state of the listener's hearing, his age, and his linguistic background. The role of the first of these, hearing, is generally understood. Of less general understanding are the subtle but demonstrable deficiencies imposed on the listener by advancing age and by having started life speaking a language other than his present tongue. These deficits in the perception of speech usually occur when the message has been degraded in quality in some way, such as by distortion or when heard in the presence of competing noises or other speech.

We are learning that under such conditions even a listener with normal hearing will show poorer understanding of speech as he advances through middle age and later than when he was in his twenties. Similarly, it appears from our recent studies that at *any* adult age the understanding of speech heard under less-than-optimal conditions will be relatively worse for one whose first-learned language was other than that to which he is listening, even though he may be completely fluent in the latter.

A reasonable question to ask is, "What features of the test language are responsible for this developing inferiority, with age, in its perception, or its persistence in apparently fluent talkers of a second language?" Some differential information is beginning to emerge from our studies and those of others. For example, it appears that if we interrupt speech a number of times per second (by processing it through an electronic switch) the effect on young adult listeners is mild, whereas the perception of many older listeners suffers near-total destruction. This, however, is not too dissimilar, in its effects, for native-born as opposed to non-native speakers of the language. On the other hand, listening to messages in the presence of a babble of other talkers is less sensitive to the aging process but clearly separates the foreign-born adult listener of any age from the native-born.

It can be demonstrated that different forms of messages, such as simple, complex, and compound, as well as the active and passive and the positive and negative varieties, are processed differently by a listener. It is appropriate to investigate whether the relative perceptibilities of such forms result in proportionately greater differences in those who begin to show changes in speech perception, as in aging. Results of modest first studies indicate that older listeners do in fact score disproportionately lower on the more difficult syntactic arrangements if they are listening under conditions of signal degradation. It has been learned further that in listening to complex sentences, particularly where the relative clause is embedded in the middle of the sentence, the older listener tends to remember only the main clause, failing to recall the relative clause, which is apparently stored while the main clause is being processed. It is reasonable to assume that similar factors operate in various degrees in all listeners.

The role of semantics in the perception of speech is somewhat more difficult to study. One approach is to manipulate the levels of probability of words or other parts of a message. This can be accomplished in several ways. Miller, Heise, and Lichten (1951) published their seminal paper showing the significance of the number of possible messages in speech perception. In their studies, when the listener was informed that the alternative choices of spoken words were only the two which he was shown, a minimum of audibility resulted in a high recognition score. When the range of alternatives involved several hundred words, however, continual increases in the intensity of the signal resulted in relatively small improvements in the listener's ability to discriminate the test words.

The level of probability of a message or of its parts may be varied also by the inclusion of more or less contextual clues. Thus a test word may be given to the subject after a nonhelpful carrier phrase, such as "The next word is *pills*," which would represent a condition of low probability, then later, on a higher level of probability, "In the hospital the nurse gives us *pills*." Such devices should help us to expose cognitive factors in the listener which contribute to his speech perception skills.

Additional factors peculiar to each listener involve such psychological attributes as the state of his vigilance, motivation, confidence, and suggestibility. Studies of the effects of these on message recognition have only just begun.

In brief, a broad-ranging study of speech perception should be concerned with the nature of the message, that is, its linguistic and acoustic characteristics, the nature of the transmitting medium and of the listening environment, and the nature of the listener's linguistic, physiological, and psychological makeup generally and specifically at the time that his perception is being evaluated. Each of these can enhance or inhibit the success of his understanding of the message.

Clearly, there is much work ahead for all of us who are interested in finding answers to questions such as those with which we began this essay. Our mutual intrusion into each other's areas of expertise is inevitable and, in fact, to be welcomed.

Arthur, I feel closer to you already!

REFERENCES

James, W. *Talks to teachers on psychology and to students on some of life's ideals.* New York: Holt, 1899.

Miller, G. A., Heise, G. A., and Lichten, W. The intelligibility of speech as a function of the context of the test materials. *Journal of Experimental Psychology*, 1951, *41*, 329–335.

Sagan, C. *The dragons of Eden.* New York: Ballantine Books, 1977.

Warren, R. M. Perceptual restoration of missing speech sounds. *Science*, 1970, *167*, 392–393.

Warren, R. M., and Warren, R. P. Auditory illusions and confusions. *Scientific American*, 1970, *223*, 30–36.

4

The Pronunciation Judgment Test, 1939-1978

An Approach to American Pronunciation

John W. Black*

Professor Emeritus, Department of Communication, the Ohio State University

1939: PRESCRIPTIVE RULES FROM AUTHORITIES

In the 1930s and earlier teachers of speech paid considerable attention to pronunciation. The popular *Speech Handbook* by Harry Barnes (1936, 1941) included a form on which teachers might grade students' speeches. One of the 11 items was *pronunciation*. This form was representative of ones that were in popular use. The accompanying textual material devoted a chapter to each of the items of the score sheet. An inference to be drawn by a student was that speech composition and delivery were of equal importance and that each was equal to the sum of its parts. A program for a theatrical production at a midwestern university often included on the production staff a "Director of Diction."

Another aspect of speech in the 1930s was a focus on testing. This was spearheaded by Howard Gilkinson and Franklin Knower. A regular feature of the annual SCA (Speech Communication Association) convention program was a section made up of papers on measurement.

*With the close assistance of Algeania Freeman (Norfolk, Virginia), Sheila M. Goff (The Ohio State University), Eui Bun Lee (Texas Southern University), Cleavonne S. Stratton (Murray State University), and Keith Young (Pennsylvania Hospital, Philadelphia).

The foregoing paragraphs suggest what seem to be the principal contributors to the construction of a pronunciation judgment test. Other influences that guided the senior author, a recent Ph.D., was the fact that he was teaching in a small Episcopalian college where prescription was more prevalent than it was in colleges of the freer churches; also his wife, who helped in the recording of the test, had studied with Gale Densmore. His specialty seemed to lie in noting recommended pronunciations.

The Pronunciation Judgment Test was recorded on discs in 1939. The nature of the test is apparent from the verbatim instructions to the four subsections, heard by the students:

1. Part 1 contains 36 words. These are broken into lists of nine words with each word spoken twice, in a paired manner. In each pair please select the preferred pronunciation and put a circle around the corresponding number on your answer sheet. Thus, if the first member of the pair is the "better," encircle *1*. If the second is the "better" encircle *2*. Listen to two examples.
2. Part 2 contains 20 words in lists of 10 words each. Each word is spoken once. Indicate whether the pronunciation of the word is "good" or "poor." Listen for misplaced accent, for omitted, reversed, or inserted sounds. If the word is well pronounced, encircle the *g;* if badly pronounced, encircle the *p*. Listen to two examples.
3. Part 3 contains six passages. Beside each passage on your answer sheet are four words. Indicate whether these words are pronounced well or badly in the passages by encircling the appropriate letter. Listen to two examples.
4. In Part 4 you hear 20 words. All the words contain more than one syllable and the accent is placed on either the first or second syllable. As you hear the words, indicate by encircling *1* or *2* where the accent occurs. Give the answer according to the way you hear the accent from this record. If the first syllable is accented, encircle *1*. If the second, encircle *2*. Listen to two examples.

The writer has commented frequently in the classroom on the ego involvement of the persons who recorded the cardinal vowels. There was the same problem in the recording of the items of the Pronunciation Judgment Test and calling some of the pronunciations *good*. The recording was made by the senior writer and his wife, each reading alternate lists of words. A preliminary form with 236 items was tried out in six colleges and universities with approximately 800 students responding (Purdue University, Kenyon College, Wabash College, Ohio State University, the University of Minnesota, and Baylor University). The distribution of scores in each school was normal. The scores were pooled, rank-ordered, and divided into upper, middle, and lower thirds. An item analysis showed that 89 of the items were responded to differently at the 0.05 level of confidence by the students whose scores fell in the top and bottom thirds of the range.[1]

Two forms of the Pronunciation Judgment Test were recorded, each of 100 items. Items included the 89 that were most discriminating and others that approached statistical significance as differentiating those who scored high and those who did not. Answers were also recorded by the same two voices that pronounced the words. Administration of either form required 13 minutes and scoring 7 minutes. The average error from self-scoring was less than 3%.

All the items of the test appeared in Thorndike's *A Teacher's Word Book of Twenty-Thousand Words Found Most Frequently and Widely in General Reading of Children and Young People* (1931). Correctness of pronunciation was set by the pronunciations listed in *Webster's New International Dictionary of the English Language,* second edition (Neilson, 1934). A requisite for a word to be included in Part 1 of the test was that it have no alternative pronunciation even in the colloquial entries in Kenyon and Knott (1944). Words with alternative pronunciations could be used in Parts 2 and 3 where a student only indicated whether or not the pronunciations he heard were acceptable.

1947: SOME FLEXIBILITY IN THE RULES

Analysis and judging of the items of the test continued until a report was made at the convention of the Speech Communication Association in Salt Lake City in 1947. The report noted that a comparison of test–retest scores yielded r-0.91 and a comparison of odd-even items also r-0.91 (corrected for length). Correlations among the different parts of the test ranged from 0.11 to 0.48 (these are shown in Table 3, p. 41). Various efforts were reported of attempts to ascertain correspondences between students' performance and judgment. One observer in an oral test of pronunciation listened to 63 students read a passage that contained 78 test items. The items were not selected from those of the Pronunciation Judgment Test. The scores correlated with the scores of the same students on the Pronunciation Judgment Test, r-0.67. The judgment also corresponded with those of an independent observer, r-0.91. In another instance, 52 of the words of Parts 1, 2, and 3 of the test were presented successively on a memory drum in 110 seconds. Students, working individually, pronounced the words speaking into a microphone. The words were ones for which there were no alternative pronunciations listed. Four judges listened to the recordings. No judge's score for an individual deviated more than five points from the average scores assigned him by the four judges; the correlations between the scores of individual judges and group judgment ranged upward from r-0.89. The mean scores assigned by the judges correlated with the scores from the Pronunciation Judgment Test, r-0.72.

The test became a routine exercise in a voice and diction class, a course in which all the authors except Lee were somewhat identified. Students' attitudes toward the test were positive—until the answers were heard from the recording. The reaction then was a vivid representation of the degree to which an individual identifies with his manner of pronouncing a word. Aside from this observation, it is important that in the succeeding decades since 1939 marked changes have occurred with regard to pronunciation in America. Variant pronunciations were published in *Webster's Third New International Dictionary of the English Language Unabridged* (Gove, 1961) and subsequent dictionaries.

CURRENT: THE SPEAKER CODIFIES THE RULES

In the face of the obstacles noted above and ignoring the probable ethnic bias of the test, four of the authors, Freeman, Goff, Singleton, and Young, teaching within the year in North Carolina Agricultural and Technical University (School 2), Ohio State University (School 3), Texas Southern University (School 4), and Hardin-Simmons University (School 1), administered both forms of the test to approximately 25 students each. Lee applied statistical tests to four null hypotheses:

1. Item tally and *t*-tests: Individual items do not differentiate between the upper and lower thirds of the students in terms of their scores on the total test.
2. There is no difference in the scores of students in the four universities in each of the four parts of the test.
3. There is no correlation between the scores of students on Form I and Form II of the test.
4. There is no correlation among the scores of students on the eight parts of the two forms of the test.

With the scores divided into two groups, top and bottom thirds, individual items differentiated groups at the 0.05 level of confidence, in the following numbers: Form I, Part 1, 11 of 36 items; Part 2, 10 of 20 items; Part 3, 10 of 24 items; Part 4, 19 of 20 items. For Form II, the numbers were, Part 1, 17; Part 2, 13; Part 3, 17; and Part 4, 18. Hypothesis 1 was overwhelmingly rejected in the instance of the items of Part 4 of both Forms I and II and was rejected in the majority of instances of the items of Parts 2 and 3. The item analysis also delineated items that were responded to in a manner that showed consensus among the students,

TABLE 1. Summaries of Analyses of Variance of the "Correct" Responses Made by Students of the Four Universities to Each Part of Each Form of the Pronunciation Judgment Test

Source of variance	df	Part 1		Part 2		Part 3		Part 4	
		ms	F	ms	F	ms	F	ms	F
Form I									
Between schools	3	92.25	7.50	.87	.17[a]	84.82	11.57	236.45	18.77
Within schools	89	12.30		5.12		7.3		12.60	
Form II									
Between schools	3	178.48	13.05	84.01	11.64	153.58	17.94	106.63	8.93
Within schools	89	13.68		7.22		8.56		11.95	

[a]Except for Form I, Part 2, all values of F are significant beyond the tabled values of significance.

that is, a proportion other than chance. For Form I, this was the case for 11 of the 36 items of Part 1, and for all of the 20 items of Part 2, for 19 of the 24 items of Part 3, and for 14 of the 20 items of Part 4. Corresponding numbers for Form II were 29, 17, 23, and 20.

Hypothesis 2 was tested by analysis of variance. The analysis is summarized in Table 1, and the accompanying means for the four schools are enumerated in Table 2. The F-ratio was highly significant, beyond the 0.01 level of confidence in the analyses of seven of the sets of scores made on the eight parts of Forms I and II. School 1 was consistently the leader in terms of the scores the students earned on the different parts of the test. School 3 was second, and Schools 2 and 4 were third. The differences among the schools were tested by both Duncan's (1955) multiple-range tests and Scheffé's (1953) procedures. The outcomes were essentially duplicates of each other. Jointly they generated 36 groupings of homogeneous scores among the four schools in the seven analyses of variance with significant F-values. Ten of these paired Schools 2 and 4; five paired and 11 isolated Schools 1 and 3, accounting for 26 of the 36 groupings and not allowing for much overlap between Schools 2, 4 and 1, 3.

School 1 is private and somewhat selective in enrollment; School 3 is a state university with open admission for instate students. All the participants from Schools 1 and 3 were white. Schools 2 and 4 are state universities with open admission and preponderantly the students are black. Hypothesis 2 was rejected. The mean values indicate that Form II of the test was easier for the students than Form I.

The tests of Hypotheses 3 and 4 are summarized in the arrays of cor-

TABLE 2. Mean "Correct" Scores (%) of the Students in Four Universities to Each Part of Each Form of the Pronunciation Judgment Test

	N	Part 1[a]	Part 2[b]	Part 3[c]	Part 4[d]	Mean
			Form I			
School 1	18	56.1	51.0	82.5	83.0	68.2
School 2	25	45.5	46.0	65.8	55.5	53.2
School 3	27	49.7	50.0	55.4	72.5	56.9
School 4	23	42.2	48.5	52.9	46.0	47.4
Mean		48.4	48.9	64.2	64.3	
			Form II			
School 1	18	68.6[e]	77.0[f]	82.5[g]	88.5[h]	79.2
School 2	25	52.2	59.0	61.3	65.0	59.4
School 3	27	62.2	67.0	64.6	75.5	67.3
School 4	23	51.9	54.0	55.4	63.5	56.2
Mean		58.7	64.3	66.0	73.1	

	Homogeneity	
	(Duncan)	(Scheffé)
[a]Schools	2–4; 2–3; 1	4–2–3; 3–1
[b]Schools	1–2–3–4 (n.s.)	4–3–2–1 (n.s.)
[c]Schools	3–4; 1–2	4–3; 2–1
[d]Schools	4–2; 1–3	4–2; 3–1
[e]Schools	4–2; 1; 3	4–2; 3–1
[f]Schools	4–2; 3; 1	4–2; 2–3; 3–1
[g]Schools	4–2; 2–3; 1	4–2–3; 1
[h]Schools	4–2; 3; 1	4–2–3; 3–1

relations in Table 3. The total scores of students in 1947 correlated, r-0.62 and in 1978, r-0.67 on the two forms of the test.

There has been an assumption in the planning and construction of the test that Parts 1, 2, and 3 would yield different measures of the same skill on the part of the judges. The obtained correlations hardly sustain this expectation. The intercorrelations in 1947 ranged, 4, 0.30–0.48 for Form I. Similar values for Form II are no longer available. In 1978, these intercorrelations ranged, r-0.07–0.24 for Form I and for Form II, 0.52–0.58.

DISCUSSION AND CONCLUSION

The transitory nature of pronunciation is, of course, well recognized. At the same time, there is an ego involvement in the pronunciation that one has learned to use. Thomas Sheridan was not able to mandate a permanent record of the pronunciation of English;[2] yet the social status of Shaw's Eliza Doolittle remains as low as when she was created in 1912.[3]

Kenyon and Knott wrote about pronunciation in *Webster's Second* (Neilson, 1934) along a line somewhere between descriptive and prescriptive admonitions:

> A standard of English pronunciation, so far as a standard may be said to exist, is the usage that now prevails among the educated and cultured people to whom the language is vernacular. (p. xxvi)

This statement was retained in Collegiate Dictionaries after the publication of *Webster's Third* and in editions of the *Collegiate* that were said

TABLE 3. Intercorrelations of Scores of the Parts and Forms of the Pronunciation Judgment Test

	Part 1	Part 2	Part 3	Part 4	Total
Form I, 1946 (N, 100)[a]					
Part 1		.46	.48	.11	.87
Part 2			.30	.29	.62
Part 3				.35	.77
Part 4					.68
Form II					.62
Form I, 1978 (N, 93)[b]					
Part 1		.18	.24	.43	.80
Part 2			.07	−.03	.35
Part 3				.01	.49
Part 4					.71
Form II, 1978 (N, 93)[b]					
Part 1		.55	.52	.46	.81
Part 2			.58	.49	.79
Part 3				.57	.82
Part 4					.79
Parts, including matching parts, Forms I and II, 1978 (N, 93)[b]					
Part 1	.55	.31	.32	.39	.50
Part 2	.26	.22	.17	−.05	.19
Part 3	.16	.12	.23	.07	.18
Part 4	.42	.47	.46	.71	.64
Total (I)	.59	.48	.51	.55	.67

[a]Any value of .25, significant at the 1% level of confidence.
[b]Any value of .27, significant at the 1% level of confidence; .21, <5%.

to be based upon it. There are echoes of Kenyon and Knott in the current *Webster's New Collegiate Dictionary* (Woolf, 1977), the eighth in the series:

> The presence of variant pronunciations indicates that not all educated speakers pronounce words the same way. A second-place variant is not to be regarded as less acceptable than the pronunciation that is given first. It may, in fact, be used by as many educated speakers as the first variant, but the requirements of the printed page. . . .
>
> A variant that is appreciably less common than the preceding variant is preceded by the word *also*.
>
> Sometimes a regional label precedes a variant (e.g., *South also* . . .).

The dominant overtone from Kenyon and Knott is a deference to the pronunciation of educated persons.

Portions of the Pronunciation Judgment Test have fared no better with the passing of time than Thomas Sheridan's dictionary. Of Part I, in which two pronunciations are heard, only 11 of the 36 items remain viable in Form I (25 in Form II), if a student accepts the current *Collegiate* as his criterion. In other words, the remaining items are now listed with variant pronunciations. In Part I, these are valid: *impotence, mischievous, comparable, puncture, orgy, combatant, irrevocable, larynx, dirigible, accessory,* and *monarchy;* these are not: *stomach, experiment, blackguard, respite, lamentable, cerebrum, chassis, eczema, forehead, Roosevelt, incidentally, vacuum, hemispheric, protein, veterinary, bestial, prelate, impious, culinary, hygienic, flaccid, foliage, zoology, gondola,* and *quay.* These are representative of the words of Parts 1 and 2 of both forms of the test.

In part 2, all the items of Form I survive and 17 of the 20 of Form II. In Part 3, 19 of the 24 items remain viable in each of the forms. Part 4 merely calls for the recognition of the stressed syllable as pronounced, recorded, and heard and does not depend on the pronunciations of *epitome, evidently, Detroit, abdomen, horizon, grimace, finance, and address.* The *right* and *wrong* responses to 6 of the 20 words of Part 4 of Form I did not differ significantly. In the terminology of "a motor theory of speech perception," perhaps the students heard the words in keeping with their own pronunciations.

Consideration was given to scoring the students' papers on the basis of the currently valid items. An examination of the papers gave no indication that such rescoring would materially alter the earlier result. For example, in the instances of 11 questionable items of Form II, the students favored the answers on the recording, not ones of the current dictionary. Also, a rescoring would seem to emphasize a prescriptive view of pronunciation.

Part 4 of the Pronunciation Judgment Test, the recognition of stress, is of particular interest. The scores of three groups of students were bimodal. Individuals either responded correctly or by chance. Here, the scores of School 1 were markedly higher than the others; the scores of the students in Schools 2 and 4 were chance on Form I and only 10–15% higher on Form II. The differences seem to reflect differences in training to recognize stress. From Table 3 one may infer that the recognition of stress is consistent behavior. The correlation between Part 4 and Forms I and II was r-0.71—although only showing 50% of the variance in common— higher than any of the other matching parts of the forms.

A descriptive view of pronunciation is with us. Even so, teachers cannot let bizarre and unlettered pronunciations go unnoticed and cannot ignore the fact that the educated person is expected to exhibit certain characteristics of pronunciation that reflect breadth of experience and classroom exchange of ideas. As Bronstein (1960) noted of American pronunciation, there are dialects, "some cultivated, some less cultivated or substandard" and "[a study of pronunciation] is a necessary part of the background of all students of language" (p. viii). If a contemporary dictionary records the variant pronunciations of educated speakers, the students—guided by teachers—should monitor their own usages, checking them against listed variants.

Several allusions have been made to the negative responses of students on hearing the answers. Reference should also be made to the pride that is taken by students who score high on the test. At one time students of acting and oral interpretation often scored 90–96%. Later a director of a university theatre commented, "We don't give pronunciation much attention any more. The emphasis is on characterization. Oh, with dialect plays pronunciation is important." In the current series of tests a teacher of a voice and diction course in a theatre department could not find the time to give his students the test. Drawing room comedy may be a setting for neither dialect nor educated speech. A test of the sort described here is a means of making an approximation of a check with a group of students simultaneously. The material of such a test, however, must be current. At the time the test was being constructed, a genial representative of the G. & C. Merriam Company agreed to take it. He listened to the answers and in a rare burst of emotion exploded, "Of course we make mistakes." This illustration has reference both to the updating of material and to the psychological topic of the identification of a person with his pronunciation.

This paper is a report of a reliable procedure for assessing the pronunciation according to the practices of yesterday. The lapse of time between yesterday and the present is difficult to measure. The dictionary is a historical document upon publication. Moreover, the reliability of an instrument that is based on a dictionary is limited to the geographical and social strata around which it was constructed. These were strained in the applications of the present test in 1978.

NOTES

1. This procedure has been largely superseded by more sophisticated ones in the United States; however, it remains current in many regions of central and southern Europe.
2. See Wise (1958) for Thomas Sheridan's avowed purpose in writing the *General Dictionary of the English Language*, 1780: "One main object of which, is, to establish a plain and permanent Standard of Pronunciation."
3. At the risk of being both boorish and paternalistic, the authors recommend a rereading of the cryptic Preface to *Pygmalion*, possibly the only brief essay Shaw ever wrote.

REFERENCES

Barnes, H. G. *Speech handbook*. Iowa City: Privately printed, 1936.
Barnes, H. G. *Speech handbook*. New York: Prentice-Hall, 1941.
Bronstein, A. J. *The pronunciation of American English*. New York: Appleton-Century-Crofts, 1960.
Duncan, D. B. The multiple range and multiple F-tests. *Biometrics*, 1955, *11*, 1–42.
Gove, P. B. (Ed.). *Webster's third new international dictionary of the English language unabridged*. Springfield, Mass.: Merriam, 1961.
Kenyon, J. S., & Knott, T. A. (Eds.). *A pronouncing dictionary of American English*. Springfield, Mass.: Merriam, 1944.
Neilson, W. A. (Ed.). *Webster's new international dictionary of the English language*. Springfield, Mass.: Merriam, 1934.
Scheffé, H. A method for judging all contrasts in the analysis of variance. *Biometrics*, 1953, *40*, 87–104.
Shaw, G. B. *Pygmalion*. London: Brentanos, 1912.
Thorndike, E. L. *A teacher's word book of twenty thousand words found most frequently and widely in general reading for children and young people*. New York: Teacher's College, Columbia University, 1931.
Wise, C. M. *Introduction to phonetics*. Englewood Cliffs, N.J.: Prentice-Hall, 1958.
Woolf, H. B. (Ed.). *Webster's new collegiate dictionary* (8th ed.). Springfield, Mass.: Merriam, 1977.

5

Surprise

Dwight Bolinger

Professor Emeritus, Department of Romance Languages and Literatures, Harvard University, and Professor Emeritus, Department of Linguistics, Stanford University

The company that a word keeps can often reveal unexpected nuances of meaning. What appears to be some kind of strange syntactic restriction turns out to be a reflection of the word's content. The verb *surprise* would seem to be usable with any thing or event that causes surprise, and yet there is a restriction that looks, at first glance, to be purely syntactic. What one is surprised at can be easily embodied in a *that* clause or a *have* infinitive, but plain infinitives may cause trouble:

(1) I was surprised that I fell.

(2) I was surprised to have fallen.

(3) ?I was surprised to fall.

(4) We were surprised that we agreed with him.

(5) We were surprised to have agreed with him.

(6) ?We were surprised to agree with him.

(7) It surprised me that I broke the vase.

(8) It surprised me to have broken the vase.

(9) ?It surprised me to break the vase

This is not because acceptable examples with plain infinitives are unusual:

(10) I was surprised to win.

(11) I was surprised to understand them.

(12) I was surprised to know that you cared.

Nor is it because every plain infinitive complement must even refer to something intrinsically surprising. Though one is not apt to say

(13) *I was surprised to come in.

there is no problem with

(14) I was surprised to come in and see you there.

As Donald Larkin (1977) points out, there can be an indefinite number of such chainings, so long as one arrives ultimately at a fact that justifies the meaning of the verb (surprise, annoyance, happiness, etc.):

(15) I was surprised to come in and sit down and look around and see you there.

(16) I was sorry to visit your mother and get all that bad news.

(17) I was mad as hell to step in the room and get slapped in the face.

And this sort of 'inferencing strategy,' as Larkin calls it, is not restricted to infinitive complementation:

(18) I was surprised when I came in and there you were!

(19) I was surprised, coming in and seeing you there.

(20) I was surprised to receive your letter and learn that Aunt Mamie was ill.

(21) I was surprised to feel a breath of wind.

In (21), the sensation is surprising only because there being a wind is surprising. (All the wrestling with logical form seems to have missed the fact that 'raising' is no more than a special case of such alignments. *I don't want to eat spinach* brings up the negative in the same way that *a hot cup of tea* brings up the *hot*—which 'really' belongs on the *tea*—and *I was surprised to come in and see you there* brings up the element that occasions the surprise.)

Since the use of *surprise* is secure with *that* clauses and *have* infinitives, it would be well to look at what they have in common, which may or may not be shared with plain infinitives.

That clauses (provided they have no modal, as in *I desire that they should leave*) presuppose INFORMATION. The information may be affirmed, doubted, denied, or emotionally reacted to, but there is always a representation of knowledge, which simply reflects the fact that a *that* clause is a proposition. We might say, therefore, that the sentence *I was surprised that he came* is somehow acceptable because the surprising fact, being in propositional form, is 'there,' a piece of information separate from the event.

The *have* infinitive does not convey a proposition, but it does something similar: by representing the action as prior, it too implies that it is 'there'—on the boards, a potential object of contemplation. In both cases there is a mental construct. The thing or event is present to the mind.[1]

Though it does not add much as corroboration, since we have already seen that the last element in a chain is sufficient to justify the use of *surprise*, there is at least no contradiction in the fact that verbs referring to the formation of mental constructs are commonplace as plain infinitives after *surprise*:

(22) I was surprised to learn that she was ill.

(23) I was surprised to be told that the plan was canceled.

(24) I was surprised to know (realize, see, hear) that you had been promoted.

But the presence of a mental construct is obviously not sufficient in itself to trigger the verb *surprise*. The content must be surprising, and one should inquire what relevance mental constructs have to that. Part of the answer comes from a semantic pattern that seems to be contradictory: something totally unexpected and very close to the experiencer gives doubtful results when expressed by a simple infinitive complement. Examples (3) and (9) have already shown this—they speak of accidents. Similarly:

(25) *I was surprised to tremble all over.

(26) *I was surprised to cough.

(27) ?I was surprised to get caught.

(28) ?I was surprised to stumble.

(29) ?I was surprised to collapse.

(30) ?The Saracens were surprised to catch fire (when they were doused with Greek fire).

As we look back on it, the Greek fire certainly took the Saracens by surprise, but it is odd to assign their immediate reaction by making them the subject of the verb. There seems to be some disturbing factor in what is close, immediate, and unexpected.

Assume that in order to be surprised one must take stock of a situation. The immediacy of the simple infinitive would then work against acceptability: surprise is a mental reaction that must appreciably FOLLOW the stimulus—there is a causal succession, a before and after. The mental construct, explicit in the *that* clause and temporally implicit in the *have* infinitive, is the basis for the reaction.

If the restriction is conceptual and not structural, we should be able to find a number of informal indications of the presence of a mental construct that would free the simple infinitive to be used with *surprise*. Try pairing (25) and (26) with the following:

(31) I was surprised to be trembling all over.

(32) I was surprised to be coughing again.

The progressive provides a basis of comparison from moment to moment—part of its meaning is that the action has emerged on the scene and is not expected to continue indefinitely. Its contrary is already in mind—another way of saying that there is a mental construct, an expectation the disappointment of which is the motive for the surprise.

Or take any number of mediating expressions that suggest the formation of some mental impression:

(33) ?I was surprised to agree.

(34) I was surprised to find myself in agreement.

(35) I was surprised to catch myself agreeing.

(36) I was surprised to end up agreeing.

(The *know-realize,* etc., verbs mentioned above also belong here.)
Or take actions which by their nature may be less immediate:

(37) ?I was surprised to thrill at her words.

(38) I was surprised to get a kick out of her words.

(39) ?I was surprised to float up (to lift off).

(40) I was surprised to be buoyed up.

(41) ?I was surprised to free myself from his clutches.

(42) I was surprised to escape from his clutches.

In the last pair, escaping is the result of a process that can be looked back upon.

Or use the tag *like that* as a test for the speaker's having objectified the action. This tag is not accented and does not mean 'in that manner' but sets the action against whatever else might have occurred at the time. I attach it to several of the examples starred or queried above:

(43) I was surprised to float up (to lift off) like that ('that any such thing happened').

(44) I was surprised to tremble all over like that.

(45) I was surprised to collapse like that.

(46) The Saracens were surprised to catch fire like that.

That way, similarly deaccented, can be tagged for the same purpose.

Or take any comment or elaboration that shows the action to have been prejudged: I modify (26) and (27):

(47) I was surprised to cough like such an idiot.

(48) I was surprised to get so ignominiously caught.

These are not manner adverbs so much as characterizations: under the circumstances (as mentally constructed), coughing was naturally idiotic and getting caught was naturally ignominious.

Finally, the prosody—with a little assist from intensifiers—may be enough to reveal that the action has been held under consideration:

(49) I was REally surPRISED to STUMble!

(50) The Saracens were surPRISED as HECK to catch FIRE!

(51) I was surPRISED out of my WITS to get CAUGHT!

It seems clear that the difficulties with the simple infinitive are not syntactic but semantic and lie in the nature of causation as speakers conceive it. As they appear to reflect the semantic makeup of the verb *surprise,* it is necessary now to look at the portion of the lexicon that structures causation in this way.

As might be expected, many other verbs signifying emotional reactions behave like *surprise* in relation to *that* clauses, *have* infinitives, and simple infinitives:

(52) I was crushed (overcome) that I had lost.

(53) I was crushed (overcome) to have lost.

(54) ?I was crushed (*overcome) to lose.

(55) I felt guilty that I had committed a crime.

(56) I felt guilty to have committed a crime.

(57) *I felt guilty to commit a crime. (Possible for a crime that is planned.)

(58) I was outraged that I had been mistreated.

(59) I was outraged to have been mistreated.

(60) ?I was outraged to be mistreated.

(61) I was ecstatic that I had won.

(62) I was ecstatic to have won.

(63) *I was ecstatic to win.

But opposing these verbs are others that offer less difficulty, despite the fact that some of them are fairly close synonyms:

(64) ?I was crushed (wretched, gloomy, dispirited, downcast, disconsolate) to fail.

(65) I was sorry (unhappy) to fail.

(66) *We were ecstatic (joyful, jubilant, gleeful) to win.

(67) We were happy (glad, tickled, overjoyed) to win.

(68) ?I feel degraded (disgraced) to accept.

(69) I feel ashamed (embarrassed) to accept.

(70) ?I was angry (outraged) to have to.

(71) I was annoyed (vexed) to have to.

(72) ?I was comforted to be there.

(73) I was reassured to be there.

(74) ?I was shaken to lose the money.

(75) I was disturbed to lose the money.

(76) ?We were paralyzed to hear the news.

(77) We were stunned to hear the news.

(78) *We were stimulated to arrive.

(79) We were excited to arrive.

Though there is a high proportion of past participles in this mix, and one might be tempted to look to some underlying structure such as *Having to*

annoyed me or *It annoyed me to have to* as an explanation of the acceptability of *I was annoyed to have to,* the examples do not divide along this
line. One can equally say *Having to overcame me,* but it does not improve
the status of **I was overcome to have to.* Whatever factor separates the
groups of verbs is shared by both adjectives and participles.

I believe that this factor is one of causation—the contrast between
EMOTIONS CAUSED and EMOTIONS PROJECTED (Bolinger, 1977, pp. 143–50).
It comes about through the tightness of the nexus between a main verb
and what it potentially subordinates. The looser the connection, the more
contingent the relationship is—given B, we have A: *John was embarrassed, to be in a place like that* (he was in an awkward place, and so he
was embarrassed). The tighter the connection, the more the main verb
becomes a GOVERNING verb—A projects itself onto B: *John was embarrassed to be in a place like that* (he disliked being there). Verbs and
adjectives are variously stereotyped as projective. When we say *I blush to
think of it* we express an attitude, not merely a contingent reaction: The
thing is not only the cause of my blushes but I am crying shame on it. I
may *shiver, quail, weep, sigh,* or *gag* to think of it, but I am not apt to
say **I sweat (flush, redden) to think of it* (though I may sweat or redden
at the thought of it and may definitely do any of these things *when* I think
of it). Some verbs have become totally projective. They express an attitude toward an event, not a reaction to it: *I regret to have inconve-
nienced you* differs in this respect from *I am heartbroken to have incon-
venienced you*—the latter describes a state resulting from the unwanted
action besides imputing heartbrokenness to it, and we can say of the subject *He is heartbroken,* whereas **He regrets* or **He is regretting* is
incomplete without a complement to describe as regrettable. A number of
adjectives are equally stereotyped as projective. *Glad* no longer refers to
a pure feeling of gladness—we cannot say **She is feeling very glad today*
as we can say *She is feeling very upbeat today.* Similarly,

(80) ?I was sorrowful to hear the news.

(80) I was sorry to hear the news.

One is glad or sorry ABOUT something: the emotions are not merely felt,
but projected on their cause. *Paul is dead, but don't be sorry* involves an
ellipsis; *Paul is dead, but don't be unhappy* does not, even though both
admit the complement *that he is dead.*

By a teleological switch, the projection may swing all the way from
cause to goal—one anticipates the emotion associated with a future cause:
I'm all excited to leave looks forward to leaving; *I'm sorry to leave, I can*

only stay an hour longer regrets a leaving still to come. The same word may serve both purposes, which suggests that it is vague as between efficient and final cause, with performance factors determining the difference: *I was embarrassed to say that* may mean that I said it with embarrassment or that I refrained from saying it because of the embarrassment it might bring *(I was embarrassed to say that, so I didn't say it)*. And the loading one way or the other is graded: *I was ashamed to tell them* is more apt to be used for what was not actually told.

With emotions that accompany unfulfilled actions, we find pure 'goal'—these may be both drives and fears. The lack of efficient cause as part of their involvement in the action can be seen in the unacceptability of *have* infinitives and indicative *that* clauses:

(82) *I was anxious that I left. (OK I was anxious that they should leave.)

(83) I was anxious to leave.

(84) They were aching (yearning, bubbling over, delirious, spoiling, wild, eager) to have (*to have had) their way.

The 180° turn from efficient to final cause is apparent in two of these terms, *eager* and *anxious*. With infinitives, *eager* has lost its original 'fierce emotion under restraint'; it is now purely forward-looking. Similarly, *anxious* with infinitives has mostly lost the meaning of 'apprehensive,' to become a virtual synonym of *eager* (the noun *anxiety* has not evolved in this way). Words referring to sanity are common here but introduce a complication: in place of efficient cause we find another nuance of causation: reason or justification for an opinion—'It is so because,' oriented to the speaker rather than to the subject:

(85) She was crazy to accept that job (she was eager to, or I judge her to be crazy because she did).

Crazy and *mad* have been relatively stereotyped in the sense of 'eager'; but there is always the possibility of using some more or less synonymous expression in the same way:

(86) She was out of her head to get (to accept) that job.

The effect of tightening the nexus shows up in all three of the situations examined so far: projection onto an existing cause, projection onto a future cause, and projection onto a 'reason':

(87) I was distressed, to hear the bad news.

(88) I was distressed to hear the bad news (it was distressing).

(89) I was suffering, to hear the bad news.

(90) I was aching to hear the good news.

(91) She's crazy, to think she can get away with that. (I regard her thinking so as a symptom of her general insanity.)

(92) She's crazy to think she can get away with that. (It is a crazy idea.)

(93) She's crazy to taste your delicious macaroons.

To be surprised is midway on the gradient. It is beyond pure contingency—a comma disjuncture, signifying mere result, is unlikely, whereas with intensified synonyms such as *flabbergasted, dumbfounded,* and *thunderstruck,* the comma is normal:

(94) ?I was surprised, to receive the award.

(95) I was flabbergasted, to receive the award.

This is to say that *to be surprised* is more projective than *to be flabbergasted.* But the projection does not extend to 'goal': we cannot be surprised at the prospect of doing something, as we may be ashamed, glad, sorry, or eager. Hence the acceptability ratings in

(96) *I was surprised to accept the job.

(97) I was surprised to be on the verge of accepting the job.

where the closest one can get to a future situation is imminence within a present one. Hence also the presence and absence of zeugma in the following:

(98) *I'm surprised and ashamed to have to go and apologize.

(99) I'm surprised and ashamed to realize that I must apologize.

(100) *I'm surprised and happy to be of service.

(101) I'm surprised and happy to have been of service.

Figure 1 shows the approximate locations of sample items in the space between reactive and projective. Possible comma disjuncture fades out as we move to the right (which is the same as to say that the chance of pure reactivity diminishes: *I was overjoyed, to get the news* is better than **I was tickled, to get the news*). Freedom to use the simple infinitive increases, admitting 'goal' (futurity) in Columns 4 and 5 and ending up with pure 'goal' in Column 6. The projectivity of the middle range is—appropriately—expressible for the most part with *at*, which reveals the reversal of direction in its contrast with *by:*

(102) I was stunned, to see him lose = I was stunned by his losing.

(103) I was stunned to see him lose = I was stunned at his losing.

This is why, by the time we get to Column 4, *by* is somewhat unusual:

(104) ?I was vexed by his losing.

(105) I was vexed at his losing.

REACTIVE ———————————————————————————————————— PROJECTIVE

1	2	3	4	5	6
suffer	be sad	blush	be vexed	regret	hasten
be stimulated	ecstatic	be surprised	annoyed	rejoice	hesitate
paralyzed	shaken	angry	unhappy	be glad	be eager
sorrowful	stunned	disturbed	over-joyed	sorry	inclined
	comforted	fascin-ated	excited	tickled	willing
	overcome	upset	embar-rassed	delighted	anxious
	exultant	alarmed		ashamed	crazy
	flabber-gasted	uneasy		afraid	reluc-tant
		disap-pointed			
that clause unlikely	*that* clause normal				*that* clause only with subjunctive
infinitive only with comma	simple infinitive restricted, comma acceptable	simple infinitive restricted, comma in-frequent	simple infinitive unrestricted, comma in-frequent	simple infinitive unrestricted, comma unacceptable	simple infinitive unrestricted, no perfect; comma unacceptable
			for infinitive and future mean-ing possible	*for* infinitive and future mean-ing normal	*for* infinitive normal, future only

FIGURE 1. The reactive-projective range.

For Column 5, the best prepositional paraphrase is a mixture of *at* and *about:*

(106) I was sorry to leave = I was sorry about leaving.

(107) I was tickled to leave = I was tickled at (about) leaving.

Column 6, being pure 'goal,' does not admit *at*, nor, for that matter, any *-ing* paraphrase (*I was willing about going* is impossible, and although *I was wild about going* is normal, it lacks the 'goal' sense of *I was wild to go* and suggests rather 'I was wild about the idea of going').

Missing from Figure 1 are the speaker-oriented cases ('reason' or 'opinion' rather than 'cause'). They are both reactive and projective and accordingly belong in the middle range, but they differ from the subject-oriented cases in not admitting *that* clauses *(You are nuts to have accepted that job, *You are nuts that you accepted that job)* and in freely admitting the simple infinitive *(You were foolish to stumble).*

Although there is a good deal of stereotyping, the reactive-projective range is open to accretions through metaphor and hyperbole. One who invents a sentence like *She was positively throbbing to get her point across* or *I was plunged into gloom that they rejected me* is not regarded as deviant. Furthermore, the picture of the six columns is neater than the reality. *Afraid,* for example, belongs somewhere between Columns 5 and 6—when it is not generic *(I'm afraid to travel)* it refers to the future: *I'm afraid to leave, but I must, and soon.* Copresence in a column does not signify that any two given items will take the same complements—obviously one is not apt to be glad and sorry about the same things, and fewer things lead to uneasiness than to disappointment.

As for *surprise,* the hypothesis that there needs to be a mental construct seems to reflect the reactive element that covers the whole middle range. For one to be *stunned, angry, embarrassed,* or *tickled,* we expect something BY which the emotion is aroused or AT which it is directed. This probably accounts for the acceptability ratings of the following:

(108) It upset me for John to be disliked.

(109) ?I was upset for John to be disliked.

(110) I was upset that John was disliked.

The *for* infinitive tends to be hypothetical, not sufficiently 'on the scene' to form the basis of a reaction—it is more appropriate for the last three columns. But the extraposition with *it* presupposes some kind of prior introduction, as does the *that* clause. On the other hand, make the reaction hypothetical and all is well:

(111) I would be upset for John to be disliked.

A final point: The effect of nexus is not limited to subordinate infinitives. We observe it also in a pair like

(112) Don't be annoyed if you lose.

(113) If you lose, don't be annoyed.

The first is more projective than the second: it tends to imply 'Don't be annoyed at losing.' The second detaches the annoyance.

I have cited the prepositions *by* and *at* as best exemplifying the reactive-projective gradient at its extremes. But the question of "prepositional rection" is a study in itself. There is a similar pairing with *from* and *of* in the following:

(114) They are tired (weary, sick) from working so much.

(115) They are tired (weary, sick) of working so much.

From refers to the instrument of the reaction, whereas *by* would refer to the agent. *Of* is projective in these contexts and is actually the underlying preposition for several of the earlier examples:

(116) Are you glad (proud, ashamed) that you said that?—Yes, I'm glad (proud, ashamed) of it.

The contrast between *over* and *about* is weaker—*over* is more reactive, *about* is more projective:

(117) Are you upset that they fired you?—Of course I'm upset about it; wouldn't you be?

Over would be correct here, but would imply cause of the feeling more than attitude toward the event. A better context for *over* is this:

(118) Why is he looking so angry?—He's worked up over the pointless firing of his colleagues.

As for *with*, it is about as projective as *at*, but its meaning of 'association' takes the edge off attitudes of disfavor: *She is annoyed with me (with what I said)* is less sharp than *She is annoyed at me (at what I said); She was furious with me* is less apt to refer to real fury than *She was furious at me. Displeased, disappointed*, and *dissatisfied with* express milder displeasure than *disgusted at* or *outraged at.* The more splenetic moods do not admit *with:* *She was mad (sore) with me.*

Using them with *upset*, which admits all the prepositions cited except *of*, we get the following gradient from reactive to projective:

(119) They were ⎱ by   at ⎧ what we
      quite upset ⎰ from   over   about   with ⎩ told them.

There is a good deal of stereotyping—if we can say *I was angry about it* there seems to be no reason why we should not be able to say ?*I was angry over it*, and if *glad of*, why not *happy of*? But the meanings of the prepositions do restrict their application on the reactive-projective scale.

NOTES

1. The category of 'mental construct' has other uses. So-called existential *there*, for example, is used to 'present to the mind' (Bolinger, 1977, pp. 90–123).

REFERENCES

Bolinger, D. *Meaning and form.* London and New York: Longman, 1977.
Larkin, D. *Complement form in Tamil and the subordinate parenthetical clause problem.* Unpublished manuscript, 1977. (Department of Linguistics, Georgetown University, Washington, D.C. 20057.)

6

Stuttering as an Expression of Inefficient Language Development

Jon Eisenson

Professor Emeritus, Department of Hearing and Speech Science, Stanford University

In the beginning there was the Word, and shortly after this beginning came the stuttered word. If we seek an answer as to the genesis of stuttering, we are probably only a little better informed today than we were in the biblical days. Then, for reasons known only to God, Moses spoke in a manner that may be described as dysfluent and very likely as stuttering.

Despite speculations over the centuries, we know very little more today about stuttering than we knew, or thought we knew, a hundred years ago. We may know more than the ancient Greeks knew about 2,000 years ago, let us say in the time of Aristotle, who may himself have been a stutterer. Aristotle believed that the stutterer had a tongue that was abnormally thick and therefore difficult to move about in the speaker's mouth. As a result, the tongue was too unwieldly to keep up with the speaker's thoughts.

What we knew or thought we knew as recently as about 25 years ago was summed up by Robert West. In an essay called "An Agnostic's Speculations About Stuttering," (1958) he summed up what he then believed he knew about stuttering. I shall quote directly from his essay because I owe at least this much of his legacy to you. West's facts are based upon his perceptions, his way of looking at things, and events. We must appreciate that when one becomes a speculator even if not an ardent believer, data tend to be screened and interpreted to fit a point of view. Subjective observations may be interpreted as if they are hard facts. As of 1958, then, we learn the following from Robert West.

THE TEN-FACT YARDSTICK

1. *Stuttering is a phenomenon of childhood.* Some diseases are phenomena of the precocious involution of specific organs of the body; they are classed as degenerative disorders. Some are phenomena of the tardy development of certain organs or the lack of such development; they may be called by analogy, agenerative. Clearly stuttering is not *degen-erative*, though it may be *agenerative*.

2. *Stuttering is more prevalent among males than among females.* (Though it usually begins at that period of life when there is little dif-ference in body patterns between males and females, stuttering appears to be at least partially sex-limited.) Those sex differences with which stuttering is associated seem not to be those directly dependent upon the mechanism of procreation, because no marked increase in stuttering appears at the onset of puberty, and because some procreating females stutter. The sex differences involved here are also distinguishable from masculo-feminine differences, since the few girls and women who stut-ter are not as a class more masculine than their sisters; and the boys who stutter are not as a class deviant in masculinity from the average of their sex. We must, therefore, consider these sexual differences more basic than those concerned with the provision of nature's economy as to which sex bears the offspring.

3. *Stuttering, or the tendency to stutter, runs in families,* whether or not it is inherited à la Mendel.

4. *Stuttering more often appears in families in which left-handedness occurs than in families characterized by uniform right-handedness.* It appears to be associated with the late acquisition of a basic preference for, or a greater skill of, one hand over the other.

5. *Stuttering appears more frequently in families in which multiple births occur than in families in which the children come one by one.*

6. *Stuttering is associated with the late acquisition of speech and with the perseveration into the second decade of the child's life of phonetic lapses that are normally found only in the first.* Stutterers also often show clumsy articulation.

7. *Stuttering rarely begins with an acute episode in the child's life, but usually is insidious in its onset.* The child is discovered to be a "stut-terer," but rarely is he seen to begin stuttering. In the stutterer's life the beginning of stuttering is as hard to date as the beginning of puberty, of adulthood, or of senility.

8. *Stuttering is a convulsive phenomenon.* The breaks in speech do not involve momentary failures of muscular tonus, but sudden spasms. In addition, the stutterer demonstrates generally heightened tonus of the striate musculature. His vocal inflections are less labile. Diadochonki-nesis of the muscles of articulation is slower generally than that of the nonstutterer, in some cases approaching that of the spastic paretic.

9. *Stuttering is reflexive psychosocially.* The fear that the patient may stutter in a given situation increases the tendency to stutter in that situation. This feared situation is usually one in which it is important—at least the patient thinks so—that he does not stutter. Examples are situations in which he talks to teacher, policeman, judge, or other special authority figures; talks before a class or group of persons; is expected to convey some specific information, such as an address, telephone number, his age, his father's name, for which there are no substitute verbal forms.

10. *Stuttering is manifest in persons who exhibit differences from the nonstutterer in certain basic physiological reactions.* Though these differences are within normal limits they appear significant, viz: The heart rhythm of the stuttering child does not regularize itself as early as does that of the nonstutterer; stutterers as a group are more allergic than nonstutterers; they show higher blood-sugar ratings than do nonstutterers.

Robert West held that the ten factors in his yardstick were "stubborn facts," each of which must be considered, explained, and incorporated by any theorist who would hold and advance a working hypothesis as to the cause of stuttering.

These facts, and others yet to be discovered, are part of the jigsaw puzzle that someday may be assembled to furnish the answer to our enigma. None of the current published theories of the etiology of stuttering furnishes a frame in which the pieces of this puzzle may be precisely and completely assembled.

So said West in 1958. The challenge still holds today.

West's observations and speculations were limited to stuttering that had its onset in childhood. He ruled out cases of persons who did not begin to stutter until they were late in adolescence or in adulthood. He judged such stuttering to belong to another syndrome, a psychoneurotic or hysterical manifestation.

It is evident that West believed stuttering to be basically organic in origin. Further, he held that stuttering probably has a single cause, but many possible precipitators. He thought it both possible and important to distinguish between the basic cause and the precipitating factors that trigger a constitutional inclination.

We may take exception with West's Ten-Fact Yardstick about stutterers. From my point of view, the observation that "stuttering is a convulsive phenomenon" is an assumption and a conclusion rather than a fact. I also have considerable reservation about the metabolic differences that West and some of his students found in stutterers when comparing them with nonstutterers. It is possible, perhaps even likely, that these differences, if present, are a result of a reaction to stuttering rather than a basic initial organic difference.

The other factors have stood the test of time. Stuttering *is* a phenomenon of childhood. Howie (1981) and Records, Kidd, and Kidd (1976) recently provided hard evidence that stuttering is biogenetically inherited. Its greater prevalence in males compared with females is indisputable. I am especially impressed with West's observation, supported by considerable research on the part of his students and associates, that stutterers are somewhat retarded in acquisition of language compared with most nonstutterers. I am particularly impressed with his observation of when, in their stage of language acquisition, children are likely to begin to stutter. Says West (p. 184):

> One-word speakers do not stutter, nor do children who murmur their sentences without any discrete words. Stuttering begins, if it does, at some time during the months that follow, as the child is learning and storing up words, parts of words, phrases, short sentences, and parts of sentences, that will make up the repertoire he will later use in an almost infinite variety of combinations for automatic use in communication. Stuttering is a failure of these first-stored neurograms to function automatically.

It is tempting to continue with an appreciative analysis of Robert West's essay. I hope that those who may not yet have been born when this essay was first published will enrich their heritage by reading the essay. But I must move on to a contemporary of West and then to my own position on early stuttering as an expression of inefficient language development in children.[1]

Bluemel (1957) in his interesting and much neglected book *The Riddle of Stuttering* explained his position and outlined his therapeutic procedures for helping the young incipient stutterer (dysfluent child) to organize language and so improve fluency of speech. Bluemel wrote as a psychiatrist and clinical theorist. He had, in fact, been a proponent of an earlier theory that stuttering (stammering) is a result of inhibition associated with speaking. The inhibition occurred because speech, viewed by Bluemel as a conditioned response, was not securely established. However naive Bluemel may have been in his views of speech and stuttering (stammering) as conditioned responses, he was ahead of his time in viewing stuttering as having its origin in early childhood language improficiency.

In the next section of this paper I will review some of the recent research on the linguistic proficiency of young stutterers. We may note that findings are by no means unanimous. Differences in population and research design do not always permit us to make valid generalizations of our observations. But the search is on, and I think that we are on the right track and are beginning to ask discerning and relevant questions.

Questions and observations can be right and appropriate only when they are addressed or related to the right persons. But children, especially those as young as two and three, are not very good at answering questions asked by adults who have forgotten how they thought and acted as children. As a result, we are inclined to undertake our investigations with adolescents and adults and make our observations on them and their stuttered and otherwise dysfluent speech. Unhappily, by doing so we can never be certain as to which may be or may have been the initiating or onsetting cause of stuttering, or which may be the maintaining causes. Some of us even make the simplistic assumption that if behavior can be modified or unlearned by operant procedures, it must have been initially established or learned by such procedures. This, from my point of view, is analogous to holding that if aspirin helps to overcome a headache then the lack of aspirin is the initial cause of the headache. To be sure, the speech and associated behaviors that identify stuttering may well be learned and therefore may, at least for a time, be modified or unlearned. But these are the surface characteristics of stuttering and tell us more about how stutterers learn to behave than why they have had to develop such behaviors.

In terms of speech as a productive act, Perkins (1977, p. 299) defines what the stutterer does in the statement "Stuttering is the abnormal timing of speech sound initiation." Under this definition are included the usual factors of hesitation, repetition, and prolongation. Perkins reviewed several investigations on speech dysfluencies (pp. 296–297). He generalized that dysfluencies associated with short units of speech—phones and syllables—are more likely to be regarded as stuttering than dysfluencies on larger units—whole words, phrases, and sentences.

Jonas, a controlled if not a cured stutterer, makes essentially the same point as to when, in his attempts at speaking, he was (is) most likely to have difficulty. In his essay *Stuttering, The Disorder of Many Theories* (1976), Jonas says:

> There were times when I got as far as the first sound in the difficult word and could do nothing but repeat it like a broken record, in the classic stutter that is imitated—usually for laughs—in books and movies. More often I had a complete block; I would try to form the first sound in the word and something within me would snap shut, so that if I opened my mouth nothing came out.

Bloodstein, in his essay "Stuttering as Tension and Fragmentation," (1975), offers much the same explanation of the moment of stuttering as does Jonas. Bloodstein is more technical and hints at a likely underyling dynamic of the stutterer's difficulty. He says:

> An especially good way to frame this concept of the moment of stuttering is to say that it is a reaction of tension or fragmentation resulting from the threat of failure in the performance of an automatic, serially ordered activity.

I prefer to restate Bloodstein's concept by modifying the last part of his statement to the effect that the *fragmentation results from experiences* and hence the apprehension of likely failure in the production of utterances in communicative situations. Most truly communicative utterances are not automatic but must be organized by the speaker for the specific situation that involves two or more speakers who are also listeners and who somehow must signal their understanding of what is said in some accepted way. The essential difference between my position and Bloodstein's is in our notion of what kind of language is "an automatic serially ordered activity." Aside from social gesture phrases, counting, saying the days of the week or the months of the year, singing or reciting well memorized content and "reading" it from memory as in an actor's or singer's performance, most of what we have to say is not preorganized for us, except as our utterances are supposed to pay some respect to the syntax of our linguistic system. It is significant that most stutterers are most fluent in their productions of the ready-made organized content. A few, who are neurotic, may even have difficulty with such content.

This brings us to the basic theme of my paper, that stuttering in its onset is a result of inefficient language development in young children. Young, for our purpose, is in the age range of two and a half to eight or nine years, children of nursery school, kindergarten, and primary school age.

This part of my paper will cite the observations of respected authorities in the field of stuttering as to onset of stuttering and language stage or level of language proficiency.

Van Riper (1971) described the stuttering of the majority of 44 children whom he followed from early onset to maturity. He refers to speech behaviors as Track II Stuttering. Van Riper makes the following points:

1. Most . . . of these children showed retarded speech development and did not use phrases or sentences until they were from three to six years of age. What is distinctive is that the disorder appeared when they began to talk consecutively. . . . The onset of stuttering came with the onset of connected speech.
2. These children are able to understand long and complex sentences for some time before they begin to communicate verbally.
3. The speech is very unorganized. Arranging, sorting, sequencing according to dimly perceived, standard syntactical patterns, these children need more than the usual amount of time and usually do not get it. They need simple models and much patience from their listeners and when these are not forthcoming, breaks in fluency appear.

Van Riper also reviewed some of the European literature and found a consensus of findings that young stutterers tended to show delay in speech development.

Andrews and Harris (1964) note that many stutterers are delayed in speech and have more defects of articulation as they acquire speech. In fact, they note (p. 108) that late talking, poor talking, and a family history of stuttering identified 120 of the 160 children in their long-term study of stutterers.

Bloodstein (1975, p. 35) notes that many stutterers are somewhat delayed in speech; many of them had not said their first words until age 1½ or 2 and their first sentences until 2½ or 3. However, Bloodstein also noted that some parents are unreliable reporters of what constitutes delayed speech and seem to have unrealistic expectations of what constitutes normal speech. For the most part, however, Bloodstein's stuttering population "corroborated past research findings of speech retardation among stutterers."

Bloodstein also notes (1975, p. 37) that there is indisputable evidence that children identified as stutterers are also defective in articulation. "Ordinary defects of articulation appear to constitute the most common single provocation to stuttering."

Bloodstein (1975, p. 57) describes the speech of a young (his Phase 1) stutterer as "doing what a child might be expected to do if he were fragmenting whole syntactic structures rather than words."

There are numerous earlier studies and observations by clinicians as to the delayed onset of speech and the higher incidence of articulatory defects among stutterers compared with children who are not identified as being either abnormally dysfluent or stutterers. These are summed up in the observation by Berry and Eisenson (1956, p. 252): "Many stutterers' histories reveal that they were retarded in the onset of speech, and maintained infantile speech patterns of articulation for longer periods than do most children."

Recent investigations of the language proficiency of stutterers provide us with support for earlier observations of clinicians as to the nature of the language improficiencies beyond deficits in articulation. I shall review several studies, especially ones that had young children as subjects.

Silverman and Williams (1967) studied 22 kindergarten and first-grade stutterers and 22 matched controls. On the basis of a 50-response language sample they found a general tendency but one not large enough to be statistically significant for measures of mean length of response, mean of the five longest responses, and structural complexity scores. A statistically significant difference was found for the number of one-word responses (approximately twice as many) than for the nonstuttering children.

Pratt (1972) in a doctoral study investigated 17 stutterers and a matched group of fluent children between the ages of 3.4 and 5.8 years.

Pratt developed an experimental battery of eight linguistic tasks to measure aspects of receptive and expressive syntactic, phonological, and vocabulary skills and "the operation of three perceptual cognitive strategies; and certain quantitative and qualitative strategies of spontaneous speech output."

Pratt found that:

1. the performance of stutterers was significantly below that of nonstutterers on all linguistic tasks except for the mean length of response and the production correlate of the actor–action–object perceptual cognitive strategy; 2. older subjects performed better on expressive and receptive syntax and vocabulary tasks than did younger subjects; 3. both production and perceptual aspects of tests concerned with vocabulary size, phonology and syntax significantly differentiated stutterers from nonstutterers; 4. the speech of nonstutterers contained significantly more sentences with multiple clauses than did the speech of stutterers; 5. stutterers produced sentences associated with the operation of semantic constraints significantly better than any of the other types of sentences. This finding indicated that this was the only cognitive strategy used perceptually by stutterers, whereas nonstutterers used two cognitive strategies, the one that operated through the use of semantic constraints, that stutterers also used, and the one in which internal syntactic relations operate on the external arrangement of words in a sentence.

Muma (1971) analyzed language samples obtained from 13 highly dysfluent four-year-old children (not identified as stutterers) and 13 matched highly fluent children. He found that the fluent group used significantly more double-base transformations[2] than did the dysfluent group.

Emrick (1972) studied the language performance of three groups of children: stutterers, nonstutterers, and highly dysfluent children who were nevertheless not regarded as stutterers. Each group consisted of five kindergarten and five first-grade boys. The subjects were required to describe pictures and respond to five parts of the *Torrance Test of Creative Thinking* (Research Edition, 1966, Revision of *Minnesota Tests of Creative Thinking*), Verbal Form A (semantic-explanation tasks). Analyses of the resulting language samples revealed that:

The three groups of children were not different from each other in the number of words in communication units (CUs), number of CUs, mean length of CUs, proportion of one-word CUs, frequency of the use of auxiliaries, transformations (negative, question, and passive infinitives), and subordinate clauses, and correct response scores on the Torrance tasks. The typical nonstuttering children, however, used more multiple word communication units (CUs) and more adjective subordinate clauses than the stuttering and highly dysfluent nonstuttering children.

The stuttering and highly dysfluent nonstuttering children obtained significantly lower vocabulary scores, made significantly more grammatical errors, and obtained significantly higher incorrect response scores on the semantic Torrance tasks than the typical nonstuttering children. The stuttering and highly dysfluent nonstuttering children were not significantly different from each other on these measures. (pp. 5509-5510)

Emrick suggests that situational (pragmatic) variables and semantic aspects of language contain more factors affecting dysfluency than does the surface structure of language.

We could continue with several other investigations that compare young stutterers with nonstutterers. However, the findings tend to be very much along the lines of the studies I have cited. The findings may be summed up thus:

1. Young stutterers are not significantly different from nonstutterers when measured by mean length of response in "spontaneous"— really elicited speech—and in the number of communicative units per response.
2. Young stutterers do differ from peer-age nonstutterers in regard to:
   a. Greater number of single-word responses.
   b. Simpler syntactic structures.
   c. Number of grammatical errors.
   d. Later onset of speech.
   e. Greater number of errors of articulation.
   f. Language at the automatic level as measured by the Illinois Test of Psycholinguistic Abilities.

Recently members of the Institute for Childhood Aphasia at San Francisco State University reported the findings of four case studies of young stutterers to the International Congress of Applied Linguistics (Eisenson, Stemach, & Lindenstein, 1978). Essentially, this was a pilot study intended to discover whether there were any trends or differential characteristics in the productive language of young stutterers. The subjects were three boys and one girl ranging in age from 5-0 to 8-8. Each child was administered the Peabody Picture Vocabulary Test to assess receptive language comprehension. Productive language was assessed on the basis of an analysis of a sample of 85 elicited utterances. The Tyack and Gottsleben (1974) procedures were used for the analysis of the samples. Descriptions and findings for each of the children follow.

CASE STUDY 1: JAMES

On the PPVT James achieved an age score of 8-11. This score is three months above his chronological age of 8-8. James presented an interesting stuttering behavior. He described each picture with almost complete fluency. However, when something from the picture reminded him of a related topic for which he was required to provide his own imagery, he became extremely dysfluent. It seems likely that, for picture description, James concentrated primarily on his language organization. When no picture was available, he was required simultaneously to concentrate on both his visual memory and language organization. James' parents speak Black English Dialect.

James achieved a MLU of 7.87. The highest language level (Level V) described by Ingram and Eisenson (1972) ranges from 5.0 to 6.0 words per sentence. One might assume, therefore, that James would demonstrate mastery of all syntactic abilities typical of a Level V child. On the basis of our sample, this was not so.

In our sample, James demonstrated inconsistent use of the prepositions *in* and *on* (Level II). Past-tense markers were correctly produced 7 of 11 times (Level V). These usages do not appear to be a function of Black Dialect.

Although James produced 28 complex sentences in this sample, these constructions were generally restricted to two types: conjoined and infinitive complement. Again, these are among the earliest acquired complex sentence types in the language of normally developing children.

Case Study 2: Ernesto

Ernesto comes from a home where Spanish is the first and English a second language. In practice, the family is bilingual, with the older members more inclined to speak Spanish than English. Ernesto's stuttering behavior was characterized by repetitions throughout the sample. He achieved a PPVT age score of 8–0. This score is seven months below his chronological age of 8–7.

Ernesto's sample produced a MLU of 5.5. This placed him at Ingram and Eisenson's Level V. Except for a single preposition substitution and omission, all of the forms (word endings and parts of speech used to connect nouns and main verbs) that appeared in this sample were used correctly.

The most striking feature of the sample related to Ernesto's complex sentence production. He used 25 complex sentences across three construction types: conjoined (all correct), infinitive complement (all correct), and adverbial ("I go to the doctor when I am sick"; 3 of 7 correct).

Case Study 3: Phillip

Phillip was the most severe stutterer of our pilot group. The child's speech was marked by repetitions, blocks, and revisions. Phillip is from a home where English is spoken as a second language. His parents are native speakers of Tagalog. Phillip achieved a PPVT age score of 5–1. This score is one month above his chronological age of 5–0.

The sample's MLU of 6.7 placed Phillip above the highest language level described by Ingram and Eisenson (1972) by a margin of 0.7.

Many of the forms required in the sentence were either substituted for another form in the same class, produced inconsistently, or completely omitted. Thus, Phillip substituted subject pronoun *he* for *she;* prepositions *in*, *of*, and *under* for *on;* and the present tense of the verb *to fall* for the irregular past tense *fell*. He demonstrated inconsistent use of the preposition *to*, copula verbs *is* and *are*, and present progressive auxiliaries *is* and *are*, as well as the progressive affix *-ing* ("The boys *are* runn*ing*."). He failed to produce the preposition *at* and helping verb *will* when he should have done so. Although he demonstrated correct use of the past-tense verb ending *-ed*, this ending marked the verb *try* (tried) in 20 of the 22 times it appeared in this sample.

Phillip attempted to produce 40 complex sentences. In 29 of the 40, he used a form of the verb *to try* as the main clause verb to begin the complex sentence transformation. In this sample, he produced complex sentences of the following types: relative, object noun (1 of 3 correct); conjoined (2 of 2 correct); infinitive complement (23 of 28 correct); adverbial (2 of 7 correct) and verb + *that* + sentence ("I know they like me," 0 of 1 correct).

CASE STUDY 4: KAREN

While Karen produced more stutterings than any other child in the study, she stuttered freely and showed no evidence of overt concern for the dysfluent behavior. Karen is from a home where Standard English is the only language spoken. She achieved a PPVT age score of 5–7. This score is one year and four months below her chronological age of 6–11.

Karen achieved a MLU of 7.26. This places her above the highest language level described by Ingram and Eisenson (1972) by a margin of 1.26.

Most of the forms that appeared in this sample were correctly produced. However, there were some important exceptions. Karen demonstrated consistent confusion of irregular past-tense verbs. Thus, she substituted *hanged* for *hung; slide* for *slid; cutted* for *cut; catched* for *caught;* and *say* for *said*. When plural nouns followed the copula verb *was*, she failed to change the copula to *were*. Karen also failed to mark future tense in the one instance that was obligatory for the construction. Finally, she substituted a form of the masculine pronoun *he* for *she* in 4 of the 13 sentences requiring the feminine pronoun.

Approximately half of Karen's sentence types were noun phrase + verb phrase ("The man sat") or noun phrase + verb phrase + noun phrase ("The man sat on a chair"). She produced 23 complex sentences, of which 4 were conjoined, 11 were infinitive complements, and 5 were relative, object noun constructions.

Our small sample and the diversity of our pilot study population do not permit us to arrive at any firm conclusions about the syntactic competence of school-aged children. However, we believe that the trends in these case studies are much along the lines and findings of other investigators. Young stuttering children produce fewer and reduced sentence types compared with peer age nonstutterers.

IMPLICATIONS OF FINDINGS

Muma (1971) suggests that a reduced MLU may indicate a compensatory behavior that allows highly dysfluent individuals to use simpler utterances or that certain aspects of syntax may cause the breakdown. Not only did the majority of the case studies exhibit a reduced MLU (Morehead and Ingram [1973] found that normally developing children acquire a MLU of 6.0 by age 4), but most showed some difficulty generating complex sentence types especially for constructions that appear later in the language of normally developing children. This observation must be tempered, however, by the lack of hard evidence to indicate when normal children actually do acquire the various complex sentence types. Moreover, it is possible that in some instances a reduced MLU (beyond the level reported by Ingram and Eisenson) implies greater rather than lesser competence for language. For example, subject deletion in the second clause of a conjoined sentence may require another rule by the child. Although the sentence "The girl will ride the bike and the boy will too" or "The girl and the boy will ride their bikes" contain fewer words than "The girl will ride the bike and the boy will ride the bike too," more grammatical rules are required to produce the shorter sentences. In general, however, as Haynes and Hood (1977) found, both Developmental Sentence Score and total words spoken in a sample of 50 complete sentences selected for analysis according to Lee's (1974) criteria increased significantly between ages 4 and 8 in a population of 30 normal children.

On the basis of the findings of our four case studies, it is evident that MLU alone is not an adequate indicator of a stuttering child's language proficiency. Analysis should include sentence types used as well as obligatory morphemes.

Broad generalizations are not in order on the basis of this limited sample. Nevertheless, the study provides evidence to suggest the potential importance of teaching a stuttering child syntactic constructions and the use of morphemes that are ordinarily expected based on stage or level of language according to MLU for young children who are neither abnormally dysfluent nor stutterers.

Now, if we assume that stutterers are not significantly different from nonstutterers in essential cognitive ability and in their intelligence as measured by standardized tests, how do we account for our findings? Are stutterers somehow intellectually different in ways not assessed by our standardized measures? Are stutterers brain-different so that how if not what they process distinguishes them from nonstutterers? Are there any ways in which we can arrive at these differences the better to understand stutterers? I think there are, even though few of our techniques and procedures are readily applicable to young children. However, let us assume that the differences we find in older children and adults are not new—post-childhood—physiological developments and not a result of reactions to maintained stuttering. Further, let us assume that the differences we find in older stutterers were there as incipient but hard-to-detect distinctions in stutterers as young children. In effect, by the time stutterers are of high school and college age, we are able to measure both manifest behavioral and possible underlying neurological processing mechanisms that were not measurable in children. I think that by looking at some of the differences as well as some of the things we do with and for stutterers to help them to be more acceptably fluent, we can get a more complete picture of them and their problems. Then we may begin to engage stutterers in therapy with insight, and I hope with greater economy and efficiency in our combined—client–therapist—efforts. For those who are curious, differences in the ways stutterers are or may be different from nonstutterers are reviewed by Van Riper (1971, pp. 335–381), Eisenson (1975, pp. 409–416), Perkins (1977, pp. 307–314), and Beech and Fransella (1968, pp. 80–103).

NOTES

1. The interested reader might wish to compare West's "Ten-Factor Yardstick" with recently established "facts" about stuttering (see Andrews, Craig, Feyer, Hoddinotts, Howie, and Neilson, 1983; Wingate, 1983).
2. Single-base transformations involve simple sentences (e.g., "we will not go") whereas double-base transformations involve complex sentences (e.g., "we heard him call"). Muma's analysis of transformational grammar is based on the model of 1965 and earlier (e.g., Chomsky, 1965).

REFERENCES

Andrews, G., & Harris, M. *The syndrome of stuttering.* London, Heinemann, 1964.
Andrews, A. Craig, A., Feyer, A. M., Hoddinott, S., Howie, P., & Neilson, M. Stuttering: A review of research findings and theories, circa 1982. *Journal of Speech and Hearing Disorders,* 1983, *48* (3), 226–246.
Beech, H. R., & Fransella, F. *Research and experiments in stuttering.* New York: Pergamon Press, 1968.
Berry, M., & Eisenson, J., *Speech disorders.* New York: Appleton-Century-Crofts, 1956.
Bloodstein, O. Stuttering as tension and fragmentation. In J. Eisenson, (Ed.), *Stuttering: A second symposium.* New York: Harper and Row, 1975.
Bluemel, C. *The riddle of stuttering.* Danville, Ill.: Interstate Publishing, 1957.
Chomsky, N. *Aspects of the theory of syntax.* Cambridge, Mass.: M.I.T. Press, 1965.
Eisenson, J. Stuttering as perseverative behavior. In J. Eisenson (Ed.), *Stuttering: A second symposium.* New York: Harper and Row, 1975.
Eisenson, J., Stemach, G., & Lindenstein, M. Unpublished, 1978.
Emrick, C. Language performance of stuttering and nonstuttering children. *Dissertation Abstracts International,* 1972, *32(9–B),* 5509–5510.
Haynes, W., & Hood, S. Language and dysfluency variables in normal speaking children

from discrete chronological age groups. *Journal of Fluency Disorders*, 1977, *2*, 57–74.

Howie, P. M. Concordance for stuttering in monozygotic and dizygotic twin pairs. *Journal of Speech and Hearing Research*, 1981, *24*, 317–321.

Ingram, D., & Eisenson, J. Therapeutic approaches. III: Establishing and developing language in congenitally aphasic children. In J. Eisenson (Ed.), *Aphasia in children*. New York: Harper and Row, 1972.

Jonas, G. *Stuttering: The disorder of many theories*. New York: Farrar, Straus & Giroux, 1976.

Lee, L. *Developmental sentence analysis*. Evanston, Ill.: Northwestern University Press, 1974.

Morehead, D., & Ingram, D. The development of base syntax in normal and linguistically deviant children. *Journal of Speech and Hearing Research*, 1973, *16*, 330–352.

Muma, J. Syntax of preschool fluent and dysfluent speech: A transformational analysis. *Journal of Speech and Hearing Research*, 1971, *14*, 428–441.

Perkins, W. *Speech pathology* (2nd ed.). St. Louis: Mosby, 1977.

Pratt, J. Comparisons of linguistic perception and production in preschool stutterers and nonstutterers. *Dissertation Abstracts International*, 1973, *34(2–B)*, 913.

Records, M. A., Kidd, K. K., & Kidd, J. K. *Stuttering among relatives of stutterers*. Address to the annual meeting of the American Speech and Hearing Association, Houston, Texas, November 1976.

Silverman, E., & Williams, D. A. A comparison of stuttering and nonstuttering children in terms of five measures of oral language development. *Journal of Communication Disorders*, 1967, *1*, 305–309.

Tyack, D., & Gottsleben, R. *Language sampling, analysis, and training*. Palo Alto, Cal.: Consulting Psychologists Press, 1974.

Van Riper, C. *The natures of stuttering*. Englewood Cliffs, N.J.: Prentice-Hall, 1971.

West, R. An agnostic's speculations about stuttering. In J. Eisenson (Ed.), *Stuttering: A symposium*. New York: Harper & Row, 1958.

Wingate, M. E. Speaking unasserted: Comments on a paper by Andrews *et al*. *Journal of Speech and Hearing Disorders*, 1983, *48*(3), 255–163.

7

Transformations—Meaning-Preserving or Text-Destroying?

Erica C. García

Department of Spanish and Portuguese, University of Leiden

When it comes to paying him suitable tribute, it is not easy to choose among Arthur J. Bronstein's commitment to his discipline, his life-long devotion to the humanities, and his keen interest in education. And yet, the very dilemma facing us shows that the division, though commonly observed, is hardly a necessary one. A unified pursuit of the three goals actually appears as distinctly worth attempting.

Take, on the linguistic side, the hotly debated issues of whether transformations preserve meaning, what is meaning, and whether grammar should or should not be sentence-bound. These are not unrelated to the recent lively interest in discourse and text grammar, in which we note a keen awareness of the fact that much of what is in a sentence can only be understood if the context in which that sentence occurs is taken into account.

Indeed, one way of appreciating the linguistic nonequivalence of transforms is to introduce into a coherent text—ideally, one with some literary merit—transformational alternatives eschewed by the author. The disruption suffered by the text can fairly be seen as a measure of the communicative nonequivalence of the transformational variants. An experiment of this type was attempted in a course on the structure of Spanish and has been more thoroughly pursued since. We will describe here one sample of this theoretico-literary-pedagogical venture.

The text chosen (Borges, 1966, p. 18) is 'coherent' at least in the sense that it deals with the same topic: it is the narrative—by the criminal mastermind—of the true facts behind the (accidental) crime which has misled the detective hero. Dandy Red Scharlach speaks:

> El primer término de la serie me fue dado por el azar. Yo *había tramado* con algunos colegas—entre ellos, Daniel Azevedo—el robo de los zafiros del Tetrarca. Azevedo nos *traicionó;* se emborrachó con el dinero que le *habíamos* adelantado y *acometió* la empresa el día antes. En el enorme hotel se perdió; hacia las dos de la mañana irrumpió en el dormitorio de Yarmolinsky. Este, acosado por el insomnio, se *había* puesto a escribir. Verosímilmente, redactaba unas notas o un artículo sobre el Nombre de Dios; había escrito ya las palabras *La primera letra del Nombre ha sido articulada.* Azevedo le *intimó* silencio; Yarmolinsky *alargó* la mano hacia el timbre que *despertaría* todas las fuerzas del hotel; Azevedo le *dio* una sola puñalada en el pecho.
>
> The first term of the series was given to me by chance. I had plotted with some colleagues—among them, Daniel Azevedo—the theft of the Tetrarch's sapphires. Azevedo betrayed us; he got drunk with the money we had given him and he undertook the venture one day too early. In the enormous hotel he got lost; toward two o'clock a.m. he burst into Yarmolinsky's bedroom. The latter, tormented by sleeplessness, had sat down to write. He was apparently preparing some notes or an article on God's Name; he had already written the words *The first letter of the Name has been articulated.* Azevedo enjoined him silence; Yarmolinsky stretched out his hand towards the bell that would awaken all the resources of the hotel. Azevedo felled him with but one dagger blow in the chest. (My gloss)[1]

This passage was rewritten so that every clause was replaced by a paraphrase, inspired in one or another of the various transformations postulated by generative grammarians. It should be stressed that this was done systematically only inasmuch as every sentence was "drawn inside out": the transformation chosen varied, depending on which would yield the smoothest Spanish alternative for the particular individual clause. It is not the case, as will be apparent below, that the text was, say, mechanically "passivized" throughout. My rewrite of the Borges original:

> El azar me dio el primer término de la serie. Algunos colegas—entre ellos, Daniel Azevedo—*habían* tramado conmigo el robo de los zafiros del Tetrarca. *Fuimos* traicionados por Azevedo: le *facilitó* la borrachera el dinero que *había* recibido de nosotros, y la empresa *fue* acometida por él el día antes. Lo *perdió* el enorme hotel; hacia las dos de la mañana *sorprendió* a Yarmolinsky en su dormitorio. El insomnio, acosándolo, lo *había* puesto a escribir. Verosímilmente, *trataba* el Nombre de Dios en unas notas o un artículo; ya *habían* sido escritas las palabras *La primera letra del Nombre ha sido articulada.* Azevedo lo *conminó* a guardar silencio; su mano se *alargó* hacia el timbre por el cual se *despertarían* todas las fuerzas del hotel; una sola puñalada de Azevedo *dio* en el pecho de Yarmolinsky.

My English gloss of rewrite:

> Chance gave me the first term of the series. Several colleagues—among them, Daniel Azevedo—had plotted with me the theft of the Tetrarch's sapphires. We were betrayed by Azevedo; the money he had received from us made drunkenness easy for him, and the venture was attempted by him one day too early. The enormous hotel got him lost; towards two a.m. he surprised Yarmolinsky in his bedroom. Sleeplessness, tormenting him, had set him to write. Apparently he was dealing with God's Name in some notes or an article; the words *The first letter of the Name has been articulated* had already been written. Azevedo ordered him to keep silent; his hand stretched out towards the bell whereby all the resources of the hotel would be wakened; a single dagger blow from Azevedo struck Yarmolinsky's chest.

It should be stressed that the preceding rewrite is totally grammatical: each sentence is unimpeachable on its own, and even the whole describes essentially the same incidents as the original. And yet . . . there must be, one would think, some difference between original and rewritten version. Or could it be that Borges hit upon the version he did write merely by chance? One would like to think that in some way Borges's original is better than the rewrite. The nature of this *better* is of interest to linguists (since it would show in what way transforms are nonequivalent), to students of literature (as suggesting one approach to the definition of good style), and to teachers of composition.

One difference between the two texts is very clear: the rewrite (certainly in the Spanish) reads more jerkily, and examination of the text reveals that this is due to a lack of continuity in focus throughout the passage. This can be shown most clearly if we ask what entities are subjects of sentences in both passages.

We find that of the 15 clauses in the Borges original, Azevedo is the subject 7 times, Yarmolinsky 4 times, and four other entities are subject once each; there are six different subjects altogether. In the rewrite there are similarly 15 clauses in all, of which Azevedo is the subject three times, Yarmolinsky once, and 11 other different entities once each, that is, 13 different subjects in all. We can conclude that the reader will necessarily have to switch focus much more often in the rewrite than in the original.

That is indeed so: if we count how many of Borges's clauses switch focus, that is, show a different subject from the one found in the preceding clause (these are italicized in the Spanish passages) we find that out of a theoretical maximum of 14 (15 minus 1, for the first clause) there are 9 switches, a 64% switch of subject. In the rewritten passage, on the other hand, 14 out of the theoretical maximum of 14 clauses show a change of subject, a 100% switch. It is clear, then, that Borges's passage is 'coherent' in a deeper sense: it focuses the reader's attention on fewer things and therefore keeps it focused on the same thing for longer stretches. If the

TABLE 1. Choice and Expression of Primary Characters

		Borges original	
Clause #	Subject	Mentioned	Second character
3	Azevedo	Yes (name)	
4	Azevedo	No	
5	first person plural		Azevedo (dative)
6	Azevedo	No	
7	Azevedo	No	
8	Azevedo	No	Yarmolinsky (prepositional phrase, name)
9	Yarmolinsky	Yes (demonstrative pronoun)	
10	Yarmolinsky	No	
11	Yarmolinsky	No	
12	Azevedo	Yes (name)	Yarmolinsky (dative)
13	Yarmolinsky	Yes (name)	
14	(Subordinate clause concerning alarm bell)		
15	Azevedo	Yes (name)	Yarmolinsky (dative)

		Rewrite	
Clause #	Subject	Mentioned	Second character
3	first person plural		Azevedo (prepositional phrase, by name)
4	*dinero*	Yes (name)	Azevedo (dative)
5	Azevedo	No	
6	*empresa*	Yes (name)	Azevedo (prepositional phrase, pronoun)
7	*hotel*	Yes (name)	Azevedo (accusative)
8	Azevedo	No	Yarmolinsky (prepositional phrase, by name)
9	*insomnio*	Yes (name)	Yarmolinsky (accusative)
10	Yarmolinsky	No	
11	*palabras*	Yes (name)	(N.B.: Yarmolinsky not mentioned)
12	Azevedo	Yes (name)	Yarmolinsky (accusative)
13	*mano*	Yes (name)	Yarmolinsky (possessive pronoun *su*)
14	(Subordinate clause concerning alarm bell)		
15	*puñalada*	Yes (name)	Azevedo and Yarmolinsky (prepositional phrase, by name)

clauses are transformed this continuity is lost, and the reader must flit from one entity to another.

Though the continuity of focus manifests itself most clearly in sameness of subject from clause to clause, it also reveals itself, more subtly, in the choice of linguistic means—nonmention, pronoun, name—with which the two principal characters (Azevedo and Yarmolinsky) are referred to. We set out in Table 1 for purposes of comparison, a reference scheme for Borges's passage and our rewrite.

It will be clear that Borges adheres to the well-known practice of introducing a character by name, then referring to him by a pronoun, then leaving him unmentioned (i.e., *he* is understood from the context). Since our rewrite was done on a strictly clause-by-clause basis, the flow of pronominal reference and of 'understood subjects,' so crucial a feature of Spanish syntax, was necessarily disrupted. Indeed, the layouts show clearly that pronominal reference is much easier to keep straight in the Borges passage than in the rewrite, where antecedents are lodged at varying distances and depths within preceding sentences.

Finally, it is worth pointing out that the choice of subject of individual clauses is not unrelated to the subject of the story. Borges has Azevedo as the subject 7 times out of 15 clauses: this is because the story he wants to tell is about how Azevedo came to kill Yarmolinsky by accident.[2] But also we are able to gather that this is what Borges is writing about precisely because Borges consistently makes Azevedo the subject about half the time. In our rewriting, however, Azevedo is subject 3 out of 15 times, Yarmolinsky only once—and the question thus arises: what is our paragraph actually about?

We conclude, then, that Borges's text is better than ours at least in being more to the point: the choice of individual linguistic means (e.g., what entity is made the subject of the individual clauses, how the various entities are referred to) is coherent with a clear and specific overall goal (to tell the story of Azevedo's crime) and it contributes to that goal. In short, Borges's writing is COMMUNICATIVELY EFFICIENT. Our rewrite is not.[3] From this we may conclude that Borges did not come to write as he did 'by chance.'

It must be apparent that an indiscriminate throwing together of grammatical sentences does not necessarily add up to coherent communication. For discourse to be coherent, the individual sentences must cooperate in transmitting a message above and beyond the one in the individual sentence. This has long been known by teachers of writing and of literature. Some linguists, however, still appear to be unaware that the properties of sentences (grammaticality aside, perhaps) may have some connection with what the speaker wishes to say: for them, this diversion may not be without value.

NOTES

1. This is a close, though not literal, translation of the Spanish original.
2. It may be pointed out that the overall coherence of the passage explains

Borges's use of a passive (a fairly unusual construction in Spanish) in the very first sentence. By this means he draws attention to *el primer término de la serie* (i.e., the first crime in the series of four); it will be noted that the whole passage is an elucidation of that first crime.

3. If we give our rewrite to native speakers of Spanish to read they are certainly able to make out what happened; they report, however, that it is "not easy to understand." It would appear that if this reaction could be quantified we would be close to a measure of internal coherence, at least to the extent that coherence is responsible for ease of understanding.

REFERENCES

Borges, J. L. *Antología personal* (2nd ed.). Buenos Aires: Editorial Sur, 1966.

8

A Method for Eliciting Verbal Graffiti

Louis J. Gerstman and Marcia Rich-Siebzehner

Department of Psychology, College of the City of New York of the City University of New York

There are three reasons why the authors are proud to contribute the following memoir to this *Festschrift* for Professor Bronstein: most relevantly, because the work comes from a doctoral dissertation by the second author (Rich-Siebzehner, 1979) that was comentored by Professor Bronstein and the first author; next, because the first author wishes to acknowledge his gratitude to Professor Bronstein for inviting him aboard the CUNY speech and hearing doctoral faculty 13 years ago; finally, because we wish to share with our colleagues yet another example of Professor Bronstein's breadth and open-mindedness regarding what research on language behavior is worth supporting. The answer, of course, is any question that possesses novelty and that can be approached with scientific rigor.

The problem in this study was to determine whether there were sex and age differences in social language use, as measured by verbosity, sentence length, and the use of expletives. To deal with this question it was decided early in the design phase that three different age groups should be employed ranging from prepubescent to adulthood (fifth grade, tenth grade, college). Additionally, at each age level, large samples should be employed in order to achieve statistical power (usually not the case in prior research) and that interpersonal factors should be avoided, such as sex of experimenter or individual differences in shyness before a microphone. For all these reasons it was decided to eschew face-to-face interviewing in favor of written responses to standard stimuli.

In these circumstances, it was felt that the best opportunity to retrieve naturalistic language would be provided by scenarios containing provocations by an actor to which subjects would be invited to make anonymous replies. The activity would thus be somewhat akin to the invitation a wall presents to a graffiti artist. It was, of course, essential that the scenarios appear equally plausible to a ten-year-old or a college student and that none of the provocations be more appropriate to one sex than the other.

Beyond these general desiderata, it was necessary that the scenarios vary in respect of our research questions, that is, the sex and age of the actor. Additionally, it seemed reasonable to determine whether it mattered if the actor spoke to the subject or acted on the subject, particularly as this might influence the use of expletives in responses.

Accordingly, after much trial and error, four scenarios were devised which met these criteria and which could be presented with either a male or female actor. Table 1 displays the final stimuli, coded according to their roles in the design and with parentheses denoting the two sex variants of each protagonist. These were produced in eight-page booklets, each containing a scenario and space for a written response. The booklets were assembled in all possible permutations subject only to the constraints that both sex versions of any scenario were adjacent and that all child scenarios (1 and 2) always preceded all adult scenarios (3 and 4).

In the dissertation a total of 225 persons completed the booklets (95 fifth-graders, 52 tenth-graders, 78 college students) all recruited from a stratified mix of public and private schools. Males and females were almost equally represented (102 vs. 123). The reader is referred to the dissertation for the full developmental findings on verbosity and sentence

TABLE 1. Four Provocative Scenarios

Scenario	Role	Text
1	Child speaking	You are standing on the corner holding two ice cream cones when a (boy/girl) on roller skates stops in front of you and says: "Why are you holding two ice cream cones?" What do you say?
2	Child acting	You are standing on a corner and a (boy/girl) on a tricycle runs over your toe. What do you say?
3	Adult speaking	A teenaged (boy/girl) meets you on the street and says: "Seeing as it's not Halloween, tell me why you're wearing a Halloween mask." What do you say?
4	Adult acting	A teenaged (boy/girl) shoves you in order to get ahead of you in line. What do you say?

length. Suffice it in this brief report to focus on the success with which the scenarios provoked naturalistic behavior, as measured by expletive use.

At the outset, we should expect that the necessity of writing one's responses would inhibit the production of expletives, and indeed 147 subjects never did so. The remaining 78 subjects produced 247 of them, which represents only 4% of their words, or about one expletive for every two scenarios. It is clear that the case for naturalness cannot depend on the quantity of expletives as such.

Conversely, a case can be made based on the demographics of expletive users and the distribution of expletives over the design factors of the scenarios. Regarding the users, tenth-graders produced expletives at twice the rates of fifth-graders or college students, and males outproduced females by 50%. Both these trends support naturalistic expectations from prior literature (Bailey & Timm, 1963; Lakoff, 1973). Regarding the scenarios, the only consistent trend over all three age groups was that adult scenarios (3 and 4) provoked twice as many expletives as did child scenarios (1 and 2), a finding eminently compatible with real world behaviors. Beyond this, the college students responded to male actors with twice as many expletives as they employed with female actors, scoring another point for scenario plausibility. Paradoxically, the matter of speaking (scenarios 1 and 3) as opposed to acting (scenarios 2 and 4) had almost no influence on expletive production, except for a modest trend in the fifth-graders where acting was more provocative than speaking. It may be that the discrimination implied by the old saw commencing "sticks and stones" tends to erode in later development.

In sum, to the extent that the inhibitions of writing were breached, the scenarios achieved their desired goal of provoking plausible responses. The reader is invited to consult the dissertation for poignant illustrations of responses ranging from extreme courtesy to intense aggression and for lists of the expletives that were elicited.

In closing, we are happy to include this study of natural language behavior on the long list of works that Professor Bronstein has fostered. On his multibranched linguistic tree we are perched at the intersect of sociolinguistics and lexicography.

REFERENCES

Bailey, L. A., & Timm, L. A. More on women's and men's expletives. *Anthropological Linguistics*, 1976, *18*, 434–448.
Lakoff, R. Language and woman's place. *Language in Society*, 1973, 2, 45–79.
Rich-Siebzehner, M. *Sex differences in the development of verbal social responses*. Doctoral dissertation, City University of New York, 1979.

9

Quo Vadunt Studia Classica?

Konrad Gries

Professor Emeritus, Department of Classical and Oriental Languages, and the English Language Institute, Queens College of the City University of New York

As is well known, the classics, that is, the study of the languages, literatures, history, and civilization of the ancient Greeks and Romans—*Klassische Altertumswissenschaft* is the apt if somewhat ponderous German name—have long since lost the preeminence which was freely accorded them during the nineteenth century and earlier. Latin, the mainstay of colonial education, the most common foreign language taught in American high schools until well after the First World War, and for decades before that the open-sesame for college entrance, has had difficulty in maintaining a place even as an elective in the secondary-school curriculum, while Greek, always recognized as being slightly esoteric—Samuel Johnson compared it to lace: "Every man gets as much of it as he can"—has all but disappeared from the public high school. In terms of numbers, the situation has, of course, affected the status of the classics at the higher levels of education also. When I was majoring in the classics at the City College of New York during the late 1920s my department occupied one of the choicest offices available: accessible, spacious, sunny; the Department of Public Speaking, as it was then called, had been allotted one that was small, rather dark, and hard to find. When I last visited my alma mater the two departments had exchanged offices. Sic transit ...!

Given the hard times on which it has fallen, CCNY is probably an unfair illustration, for there are still institutions which maintain flourishing classics departments. Even there, however, among students as well as faculty and administration, the classics are likely to be regarded a survival, a subject with no immediate, practical value and a discipline which has reached a dead end. What discoveries are left for the devotees of two dead languages and literatures to make? Have not all the classical texts been edited and analyzed ad nauseam? What possibilities for productive scholarship can there be? It is the purpose of this brief account to dispel some of the misunderstandings which are responsible for such questions.

To begin with definitions, the classicist is no longer the gerund-grinding pedant that he all too often used to be. His interests—at every level of the educational system—have broadened to include all facets of the life of classical antiquity. The college and university departments that once taught Greek and Latin exclusively are now offering courses in classical mythology, Roman law, ancient technology, even word building and derivation, not to mention the ubiquitous courses in literature in translation. At the secondary level, the teacher no longer confines himself to what I once called "the three greats": Caesar, Cicero, and Vergil; high-school Latin textbooks now introduce the student to the entire range of the literature from Plautus to the Carmina Burana, with instruction focused not on grammatical analysis but on literary appreciation and an understanding of the cultural and historical backgrounds. And the once aloof academician has now cast his net to encompass the lowly fifth-grader: he has involved himself with FLES (Foreign Languages in the Elementary School) programs in a successful effort to extend the basic values of Latin to the underprivileged children of the inner-city elementary school in such communities as Indianapolis, Philadelphia, Los Angeles, and Washington, D.C.

For the professional educationist, the introduction of Latin into the elementary school, together with the realization that nineteenth-century methods of teaching Latin (and Greek) were no longer productive—if they ever had been—of the benefits to be expected from a classical education, means the opportunity to devise new approaches and to experiment with new methods. Language laboratories, structural linguistics, the classroom use of the spoken language, individualized learning, programmed instruction—these and other innovations have transformed the Latin classroom and apparently—together with the popular call for a return to 'basics'—occasioned a minor upswing in the demand for Latin as a school subject. In any case, the Latin teacher, faced with the challenge of a utilitarian "I'm from Missouri, you've got to show me" era, is engrossed in evolving and testing new theories and novel methods. The result may not always be progress, but the activity is there.

But what of the scholar, the professor who is supposed to engage in the search for knowledge, to expand the boundaries of his chosen field or deepen appreciation of its subject matter? The most obvious and best known affirmation of the continuing need for research in the classics is provided by the archaeologists, the results of whose digs are occasionally spectacular enough to find their way into the headlines. I mention only the two most recent such discoveries. First, within the last fifteen years excavations on the Aegean island of Santorini (ancient Thera) have revealed the flourishing Minoan civilization destroyed by the cataclysmic volcanic eruption that literally blew the island apart at some time in the mid-fifteenth century B.C. and have aided in explaining the mysterious decline of that same civilization on Crete—a civilization the discovery of which by Sir Arthur Evans is itself less than a century old. Second, only a few years ago two untouched royal tombs were found and excavated in the Macedonian village of Vergina; their excavator is quite certain that one is the burial place of King Philip II, the father of Alexander the Great. Indeed, much work awaits the shovel and the spade: in unexplored areas of Asiatic Turkey, which was once crowded with prosperous Greek and Greco-Roman communities; in the southern plains of Russia; in the mountain valleys of the Balkans—and these are only the most obvious places.

The sands of Egypt, too, still hide treasures on papyrus, not to mention the wealth of documents already unearthed and waiting to be deciphered and translated. The work of the papyrologist is indispensable for the modern historian of ancient classical civilization, for the finds represent just the stuff which he needs in order to reconstruct the life of antiquity: private letters; public and legal documents of all kinds; scientific, medical, astrological, magical, and religious as well as literary texts. The latter offer not only copies of well-known masterpieces such as the *Iliad* and the *Odyssey:* from time to time works are unearthed of which the world knew only their reputation or which were entirely unknown: Aristotle's *Constitution of Athens* (1890); some 900 lines by an anonymous Greek historian, the so-called *Hellenica Oxyrhynchia* (1906); large fragments of the *Ichneutae*, a satyr-play by Sophocles (1912); and a complete comedy, the *Dyskolos*, by Menander, a fourth-century writer of New Comedy whose highly praised plays had been known only in fragmentary form (1958). The same account of continuing discovery can be given of the science of epigraphy, with new inscriptions constantly turning up. Though these are of necessity normally brief, one can point as an exception to the extensive document, considerable fragments of which were found in the 1880s and 1890s in Lycia, set up by one Diogenes of Oenoanda in the second century A.D. to proclaim the philosophy of Epicurus; additional fragments have come to light within the past decade.

When it comes to the primary task of the classical philologist, that of determining the most accurate text of a Greek or Roman author, it is indeed true that, barring unforeseen discoveries such as those mentioned above, new editions of classics like Homer, the Greek tragedians, Plato and Aristotle, Caesar and Cicero, and Horace and Vergil can offer little more than minor improvements. But classical literature consists of more than the acknowledged masterpieces of the Hellenic and Hellenistic periods and of the Roman Republic and Empire. There are extant minor works of value to the agronomist, the grammarian, the historians of jurisprudence and medicine, the rhetorician, the theologian. And of course Latin and Greek did not cease to be used by poets and scholars just because the Roman Empire "fell"; more and more attention is being paid to the Greek and Latin writers of late antiquity, of the Middle Ages, and of the Renaissance. Thus, in recent years there have been editions and/or translations of, among others, an astrological poem by Dorotheus of Sidon (first century A.D.), known mainly through an Arabic translation made about 800; a manual on husbandry and veterinary medicine by Palladius, a knowledgeable agriculturist of the fourth century A.D.; the theological writings and commentaries of C. Marius Victorinus, a Neoplatonist of the same period; a philosophical treatise by Isaak Sebastokrator, possibly a relative of the Byzantine emperor Alexius I (1081–1118); Peter the Deacon's chronicle of Monte Cassino (twelfth century A.D.); the *Ecerinis* of Albertino Mussato (1261–1329), the first humanistic tragedy; the letters, poems, and miscellaneous writings of the Byzantine humanist John Chortasmenos (c. 1370–c.1436); the 1449 *De Machinis,* an engineering treatise by the Sienese Mariano Taccola; several political works by the Bohemian humanist and bibliophile Bohuslav Hassenstein (c.1461–1510); Giovanni Armonio's *Stephanium,* the first successful Neo-Latin comedy in verse (c. 1500 A.D.); the otherwise unknown John Joncre's Latin tragedy about the Polish king Boleslav II (sixteenth century A.D.); and several seventeenth-century Latin plays written for presentation in the Jesuit College at Poznan. It is also not amiss to point to the continuing work on the scholia (marginal critical and expository notes preserved in the manuscripts) to such ancient authors as Homer, Aeschylus, Aristophanes, Horace, Vergil, and Juvenal.

There may be little work remaining to be done on the classical texts *per se,* but it is certainly true that, as Anatole France once wrote, "Chaque génération imagine à nouveau les chefs-d'oeuvre antiques." To inform and modernize this ever-changing picture of the classics requires the production of contemporary translations and modern commentaries and interpretations to enlarge, intensify, and thus alter our appreciation of them. Of the former group, called forth perhaps in part by the popularity of courses in ancient literature in translation, there seems to be no end—especially, of course, of the better-known figures: Homer, the Greek playwrights, Plato, Vergil and Horace, Seneca, St. Augustine; the translators include not only professional classicists but also gifted poets and litterateurs: Robert Graves, for instance, Jane Lembke, Aubrey de Sélincourt, Rex Warner, and Peter Whigham. As for reinterpretation, it is Homer who has been most affected—by the conclusive proof, provided by Milman Parry in the late 1920s and early 1930s, that the epic poems which go under that name are essentially oral compositions, a position confirmed by the evidence of comparative studies, especially those involving the South Slavic epic. This confirmation and the resulting change in our understanding of Homer's art has occupied Homeric scholars ever since. The interpretation of Vergil's *Aeneid,* too, has undergone drastic change, the poet no longer being regarded as a subservient yes man to the Emperor Augustus, but as a critical spirit deeply aware of the inadequacies of contemporary imperial ideals. Vergil's *Eclogues,* the elegies of Tibullus and Propertius, and the *Odes* of Horace also are being subjected to renewed scrutiny in line with the methods advocated and applied by critics of modern poetry—not always, in my opinion, appropriately or successfully. As a final example, it is only now that the aims and methods of the historian Livy are being fully understood and appreciated.

So far I have discussed developments in some of the areas traditional to the classicist: art and archaeology, papyrology and epigraphy, textual and literary criticism, translation. I conclude by calling attention to three areas, interest in which is either new to the profession or else has been made possible only by the advances of modern science. The investigation of the *Nachleben* of classical authors and antiquity, the study of *The Classical Tradition* (to use the title of Gilbert Highet's classic book on the subject), has received renewed impetus through the recent celebration of the nation's bicentennial: American classicists have turned to investigating the use made of their classical heritage by the early settlers, the colonials of the eighteenth century, (particularly the Founding Fathers), and the first citizens of the United States. There has even been a reissue of a forgotten modern Latin biography of George Washington, first published in 1835.

One of the most characteristic inventions of contemporary technology is the computer. Classicists, with a greater or lesser degree of enthusiasm, have availed themselves of the speed and accuracy made possible by this device and are applying it assiduously to such pursuits as stylometry; the metrical analysis of classical poets; the preparation of concordances to authors like Aeschylus, Hesiod, Livy, and Lucretius; the compiling of Greek and Latin dictionaries, notably the California-based *Thesaurus*

Linguae Graecae; and the establishment, by the American Philological Association, of a Repository of Greek and Latin Texts in Machine-Readable Form. Also available to classicists concerned with the application of computing to their discipline is a new journal, *Calculi,* which serves as a clearinghouse for information on pertinent projects, meetings, and publications. (It is appropriate to mention here the increasing use of computer-assisted instruction in heavily populated college courses that require much drill and memorization.)

Finally, an entirely new area of activity was created by the discovery in 1952, by the British Michael Ventris, that Linear B, a hitherto undeciphered script known originally from tablets found at Cnossos in Crete, but now represented also on mainland Greece and dating from the fourteenth and thirteenth centuries B.C., is a notation for an archaic form of the Greek language. (The related Linear A script still baffles cryptanalysts.) Around this discovery there has developed what may fairly be called a new breed of classicists, the Mycenologists, as they refer to themselves. The term derives from the fact that Linear B was used for the day-to-day accounts and inventories kept in the royal palaces of Mycenaean Greece, the setting for the *Iliad* and the *Odyssey,* but the interests of the new specialty include the entire area of early Aegean scripts, languages, archaeology, art, history, and civilization. A broad spectrum, characteristic of the all-encompassing nature of classical philology. In itself, it should suffice to convince the skeptic that the classics have by no means reached that bruited dead end. There's still gold in them thar hills, with plenty of eager and able miners to bring it to the light.

10

On Consonants and Syllable Boundaries

Katherine S. Harris

The Ph.D. Program in Speech and Hearing Sciences, the Graduate School of the City University of New York, and Haskins Laboratories

Fredericka Bell-Berti

St. Johns University and Haskins Laboratories

Arthur Bronstein, in his book *The Pronounciation of American English* (1960), follows the convention of dividing the sounds of the language into two classes—the consonants and the vowels. Within this rubric, he assigns the glottal stop [?], and the glottal fricative [h] to the consonant class, as other textbook authors do. To choose a few examples, [?] is described as a "glottal plosive" and [h] as a "breathed glottal fricative" by Daniel Jones (1956); [?] as a "laryngeal stop" and [h] as a "laryngeal open consonant" by Heffner (1949). The authors thus make the tacit assumption that these sounds share some property with the stops and fricatives and contrast, in some manner, with vowels. In part, this view is a consequence of their distributional properties (Andresen, 1968) and, indeed, their role in the syllable. However, this decision leaves us with the further problem of deciding what syllables are, within which the consonants and vowels

This work was supported by NINCDS Grants NS-13617 and NS-13870, and BRS Grant RR-05596 to Haskins Laboratories.

89

may have roles. To continue with our sampling of phonetics texts, we find Malmberg (1963) and MacKay (1978) observing that, although phoneticians may differ on the definition of a syllable, the untrained speaker of a language usually has a clear idea of the number of syllables in an utterance, and this intuitive reality suggests that there must be some corresponding articulatory reality. For convenience, we will ignore the problems of the more general definitions of the syllable (Bell & Hooper, 1978; Pulgram, 1970), though we note that the problem of finding articulatory meaning for the syllable is made more acute by the failure of efforts to find easy distributional definitions.

Modern physiological research on the syllable begins with the work of R. H. Stetson (1951), who suggested that the syllable was physiologically defined by an initiating and a terminating burst of activity from the muscles of the chest wall, the internal and external intercostal muscles, resulting in a distinct chest pulse for each syllable. This attractive concept was effectively torpedoed by the classic experimental work of Ladefoged and his colleagues (Ladefoged, 1967), who were able to show that there were not discrete bursts of muscle activity corresponding to individual syllables and, indeed, that the manner of interaction of muscular and nonmuscular forces in the expiratory cycle made the idea of a syllable based on separate muscular syllable pulses theoretically implausible. More recently, attempts have been made to salvage the concept of an articulatory syllable by assuming that its boundaries may be discovered by careful examination of the activity of the upper articulators, rather than the respiratory muscles.

Many current theories stem from the work of Kozhevnikov and Chistovich (1965), originators of the concept of the articulatory syllable, defined by coarticulation. In brief, they suggested that all elements in a single syllable are coproduced. As a consequence, for example, if a syllable contains a rounded vowel, the consonants associated with the syllable would be likely to take on "rounding" attributes. As a correlate, one might suppose that in sequences of an unrounded-vowel syllable, followed by a rounded-vowel syllable, an examination of rounding characteristics of the intervocalic consonants might permit the specification of a syllable boundary. In fact, Kozhevnikov and Chistovich suggest an "articulatory" syllable consisting of a vowel and its preceding consonant string. This basic suggestion has been amplified by Gay, who finds that in a VCV string, the articulatory movement toward a second vowel begins at, but never earlier than, the onset of the first intervocalic consonant (Gay, 1978); in other words, the syllable boundary is marked in coarticulatory terms.

Support has been provided for this idea by the so-called "trough phenomenon" (Bell-Berti & Harris, 1974; Gay, 1975). Briefly, it has been shown that if two rounded vowels of the same phonetic specification are produced in sequence, with a single consonant or string of consonants unspecified for rounding between the vowels, as in [utu], the lip muscles will relax between the two vowels, so that the consonants are produced with only partly rounded lips. The same phenomenon can be demonstrated, as well, in sequences like [ipi], where the tongue, which must be raised and fronted for the two identical front vowels, relaxes in association with production of the [p], although the conventional or feature description of [p] does not specify a tongue position for the consonant. In both cases, there are two "vowel" gestures, one, apparently, for each syllable. However, for reasons of economy of production, one might expect a "held" gesture for the second of the two vowels, since the production of the intervening consonantal gesture does not appear to be in conflict with the vowel.

While these facts can be used to argue against some models of coarticulation (Bell-Berti & Harris, 1981), they provide support for coarticulatory marking of syllable boundaries if a trough, indicating a consonant gesture, is formed at all syllable boundaries. In the textbook descriptions of phonetic sequences we provided earlier, we understood that a syllable boundary must occur somewhere in the sequence VCV. The trough phenomenon provides evidence of boundary marking because a vowel-to-vowel gesture, which might, apparently, be produced continuously, is not. If [h] and [?] are consonants, they should interrupt a vowel-to-vowel sequence in the same way as [t] production interrupts vowel rounding.

The general hypothesis is that the "trough" phenomenon is a general syllable boundary marker. We wanted to examine [h] and [?] for the two syllable sequences where the original observations of the trough phenomenon were made. We ask, Do [h] and [?] cause relaxation of the tongue for [i] sequences, and Do [h] and [?] cause relaxation of lip protrusion for [u] or [ɔ] sequences?

At present, the most effective way of observing the movements of the tongue is in lateral view cineradiography. We have made extensive observations of tongue movements using a special-purpose setup, the x-ray microbeam installation at the University of Tokyo (see Kiritani, Itoh, & Fujimura, 1975).[1] For the purposes of the present discussion, we merely noted that the output of the system is a series of plots of the x and y coordinates of the position of pellets affixed to the articulators. The speaker was a male native of New York State, with no pronounced speech defects.

Katherine S. Harris and Fredericka Bell-Berti

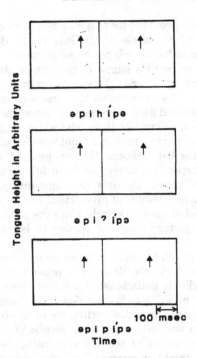

FIGURE 1. X-ray microbeam traces for the syllables [əpipípə], [əpihípə] and [əpiʔípə]. The plots show the vertical coordinate of a pellet on the tongue blade and mid-tongue position. Coordinate values with larger *y* values show greater tongue height. The long vertical line on each trace shows the time of the end of voicing for the first [i]. The two upward-pointing arrows show the beginning and end of the two-vowel sequence.

Figure 1 shows the position of the *y* coordinate for two pellets as a function of time for three nonsense syllable sequences, [əpihípə], [əpiʔípə] and [əpipípə]. An examination of these three tokens, and others like them that vary in stress and speaking rate, leads to the general impression that a trough is substantially less likely for [ʔ] and [h] than [p]; some samples of [h] show a trough, but most do not. Of course, more quantitative observations are necessary.

It is somewhat easier to observe the movement of the lips in the production of rounded vowels. Although it is possible to use x-ray methods, an easier technique is to observe the forward protrusion of the lips in rounded vowel production either directly by monitoring movies of the lips in profile, or indirectly by recording the output of a suitably placed strain gauge.

Figure 2 shows the lip movement for the sequences [lɔʔɔl] and [lɔtɔl]. Unfortunately, we did not examine the sequence [lɔhɔl]. The speaker was a female native of the Washington, D.C., area, with normal articulation. The recording shows the output of a strain gauge placed on the lower lip in such a way that forward movement of the lip causes bending of the plate (Abbs & Gilbert, 1973).[2] An examination of the figure suggests that there is a trough in the lip-protrusion curve for [t], but not for [ʔ].

Unfortunately, as with many experimental facts, the results just described may be interpreted in several ways which are not mutually exclusive. One possibility is that there is no coarticulatory definition of the syllable boundary. A second possibility is that the "laryngeal" stops [h] and [ʔ] do not form a class with [t] and [p] so that [h] and [ʔ] are not "true" consonants and thus cannot lead to boundaries even if [VhV] and [VʔV] are judged to be disyllabic. A third possibility is that existence of

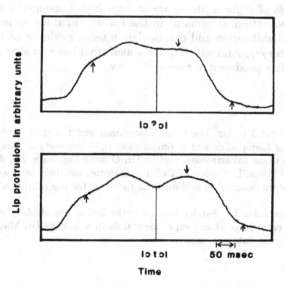

FIGURE 2. Output of a strain gauge transducer on the lower lip for the syllables [lɔʔɔl] and [lɔtɔl]. The trace shows the forward movement of the lips for rounding during vowel production. Coordinate values increase for greater forward movement of the lip. Line and arrows indicate the same acoustic events as in Figure 1.

a trough is some sort of a positive articulatory requirement for each phone for which it occurs. Such an approach is taken by Engstrand (1981); he suggests that the lip relaxation associated with [s] and [t] between rounded vowels may arise as a consequence of the aerodynamic prerequisites of these consonant sound types, rather than as a consequence of some general consonant property, or their syllabic position. Presumably, then, by analogy, lip relaxation fails to occur for sequences in which a glottal stop occupies the intervocalic position, because there is no acoustic requirement for such a maneuver. If the argument is accepted, we must then search for those acoustic requirements which specify the details of tongue position for a bilabial stop, in the environment of high front vowels. While it may seem, on the face of it, somewhat unparsimonious to search for two separate acoustic arguments for the appearance of a trough in the two environments, there is no *a priori* reason to discard the possible explanation.

Observations like those of this experiment substantially restrict the field over which we can apply any "theory" of coarticulation, or of syllabification. Nonetheless, we have ample evidence that the articulatory requirements of a given phone are at least broad enough to allow some contextual variation. It remains in the future, then, for us to develop a theory of syllabification and coarticulation using evidence gathered from the articulatory domain with a net of a mesh that has a smaller gauge than that which has produced our present views.

NOTES

1. We are grateful to Dr. Masayuki Sawashima and the staff of the Research Institute of Logopedics and Phoniatrics at the University of Tokyo for their help with these experiments, and to Dr. Osamu Fujimura, Dr. Joan Miller, and Dr. Winston L. Nelson of Bell Laboratories for their help with the preparatory data processing and analysis facilities for the output of the Tokyo System.
2. We are grateful to Dr. Sandra Hamlet of the University of Maryland for making these recordings at her experimental facility, and to Dr. Maureen Stone for her help in data analysis.

REFERENCES

Abbs, J. H., Gilbert, B. N., A strain gauge transduction system for lip and jaw motion in two dimensions: Design criteria and calibration data. *Journal of Speech and Hearing Research*, 1973, *16*, 248–256.

Andresen, B. S. *Pre-glottalization in English standard pronounciation*. Oslo: Norwegian University Press, 1968.

Bell, A., Hooper, J. B. Issues and evidence in syllabic phonology. In A. Bell & J. B. Hooper (Eds.), *Syllables and segments*. New York: Elsevier North-Holland, 1978.

Bell-Berti, F., & Harris, K. S. More on the motor organization of speech gestures. *Haskins Laboratories status report on speech research*, 1974, *SR-37/38*, 73–78.

Bell-Berti, F., and Harris, K. S. A temporal model of speech production. *Phonetica*, 1981, *38*, 9–20.

Bronstein, A. J. *The pronunciation of American English*. Englewood Cliffs, N. J.: Prentice-Hall, 1960.

Engstrand, O. Acoustic constraints of invariant input representation? An experimental study of selected articulatory movements and targets. *RUUL* (Reports from Uppsala University Department of Linguistics), 1981, *7*, 67–94.

Gay, T. Some electromyographic measures of coarticulation in VCV utterances. *Haskings Laboratories status report on speech research*, 1975, *SR-44*, 137–146.

Gay, T. Articulatory units: Segments or syllables? In A. Bell & J. B. Hooper (Eds.), *Syllables and segments*. New York: Elsevier North-Holland, 1978.

Heffner, R-M. S. *General phonetics*. Madison, Wisconsin: University of Wisconsin Press, 1949.

Jones, D. *An outline of English phonetics* (8th ed.). New York: Dutton, 1956.

Kiritani, S., Itoh, K., & Fujimura, O. Tongue-pellet tracking by a computer-controlled x-ray microbeam system. *Journal of the Acoustical Society of America*, 1975, *57*, 1516–1520.

Kozhevnikov, R. V., Chistovich, L. *Speech: Articulation and perception*. (Translated from Russian) Report 30, 543. Washington, D.C.: Joint Publications Research Service, 1965.

Ladefoged, P. *Three areas of experimental phonetics*. London: Oxford University Press, 1967.

MacKay, I. *Introducing practical phonetics*. Boston: Little Brown, 1978.

Malmberg, B. *Phonetics*. New York: Dover, 1963.

Pulgram, E. *Syllable, word, nexus, cursus*. The Hague: Mouton, 1970.

Stetson, R. H. *Motor phonetics: A study of speech movements in action* (2nd ed.). Amsterdam: North-Holland, 1951.

11

The Continuing Education of the Professional

Irving Hochberg

The Ph.D. Program in Speech and Hearing Sciences, the Graduate School of the City University of New York

The education of professional personnel has emerged as an important and continuing activity in many disciplines. The concept is based on the premise that formal academic preparation of individuals can provide knowledge that is limited to a particular time frame and cannot meet all of the needs the professional may have beyond that point in time. As new information emerges, it must be incorporated within the professional's capabilities and role function so that he maintains and even transcends his standard of performance. The assumption underlying this notion is that formal, academic education, no matter how excellent, can go just so far in personnel preparation and no further. That the education of the professional must continue beyond his academic preparation is presumed.

At the present time there appears to be a deluge of so-called continuing professional educational activities in virtually every profession. These typically take the form of short-term workshops, seminars, institutes, and the like. They usually last for no more than two or three days, attempt to cover a limited number of topics, and are designed to update the individual's knowledge so that he can keep abreast of new and developing research and information. Their short-term format ordinarily precludes intensive interaction. In fact, the nature and format of these activities reflect their ostensible purpose, namely, to provide limited information on selected topics in a short period of time, no more, no less. The success or failure of a given activity should, of course, be judged in relation to its objectives and how well these objectives have been achieved. Once it is understood that these short-term activities can only accomplish short-term objectives, a more accurate assessment can be obtained. And for what they are intended to do, they can be designed to achieve a considerable amount of success for updating professional personnel in selected areas of research and information. From a practical point of view, they also meet the educational needs of the busy professional, permitting him to maintain a level of currency in his area of interest.

These so-called short-term professional activities should not be confused with those activities which are specifically designed to achieve long-term goals, and are usually carried out in a given employment setting for the professional staff of that institution. Such activity is usually referred to as inservice training, and although the same material may be covered in short-term modules comparable to that discussed above, inservice training has the potential advantage of being long-term and can be designed to achieve quite different objectives.

For example, whereas a two- or three-day conference precludes the assessment of a teaching strategy, or the evaluation of a particular instrument with certain groups of individuals, inservice training can accommodate such activities by designing a six-month program which encompasses weekly or bi-weekly sessions. It can utilize the personnel and resources of the institution over protracted periods of time and can supplement these with outside consultants. Selected objectives for the development of particular skills can be identified, and inservice training programs can be designed for staff development.

We have, then, two distinct professional educational activities, each being defined according to its purpose, capabilities, and goals. One is not a substitute for the other, although it can be argued that if one has no other alternative than to participate in a one-day conference, that is better than nothing. Some may argue with this point as well. Nevertheless, the two types of activity would appear to be complementary in achieving the overall objective of professional education, namely, to improve individual competency. Within this framework, then, short-term acquisitions of knowledge can be integrated in an ongoing program of inservice training which may span many months.

Table 1 illustrates the conceptualization that knowledge and the development of skills lead to competency. The means by which each component may be achieved are listed with examples of corresponding activities. Knowledge may be achieved in a variety of ways on a short-term basis and then incorporated into activities such as problem solving, dem-

TABLE 1. Integration of Continuing Professional Education and Inservice Training Activites for Maximizing Competency

Knowledge	+	Skills development	=	Competency
Short-term continuing professional education	+	Long-term inservice training	=	Modification of role function
Workshop		Problem solving		Experimental teaching
Seminars		Demonstrations		Materials development
Symposia	+	Consultant services	=	Evaluation and testing
Institutes		Microteaching		Innovative stategies

onstrations of teaching, utilization of consultants' services, and the like within an ongoing inservice training program designed to develop particular skills. These activities may be continually refined and modified as new information is acquired and applied. The developing skills will thereby reflect the current state of knowledge. As newly acquired (or modified) skills emerge, they will have significant impact on the individual's role function and serve to modify his professional activities. By employing these skills, the professional may realize a more experimental and innovative style of teaching, a willingness to develop new teaching materials, and an enhanced capability to evaluate intervention strategies. In short, he may become a more dynamic and resourceful educator.

As with any approach, certain assumptions have been made. Given that any set of circumstances is characterized by any number of assumptions, the thesis being advocated here is that in the best of all possible professional worlds (and there may be one), a series of short-term educational activities integrated into a program of inservice training would appear to have the best chance of making significant and lasting impact on improving professional competency.

12

Some Data on Second Language Acquisition and Retention by Older Children

Edith Trager Johnson

Professor Emerita, Linguistics Program, San José State University

During the anxious days of the Six Day War in Israel in 1967, concerned colleagues and family members asked then and for months afterwards if I were truly going to go through with being a Fulbright Lecturer at Tel Aviv University. They were especially worried because I was going as a single parent with three young sons. Let me hasten to say that I went and that we all found it the most rewarding year of our lives up to that point.

One highlight of my career there was the happy accident of Arthur and Elsa Bronstein's living just down the street and of my being Professor Bronstein's colleague in the Linguistics Program at Tel Aviv University. I had known of him, of course, as one of those rare scholars who combine facility in the fields of both linguistics and speech with a lengthy publication record. I was also most impressed to meet the author of *The Pronunciation of American English* (Bronstein, 1960). The great usefulness and wide impact of this book on speech students throughout the country was already established. Now we were to share in teaching Israeli students phonology and other aspects of linguistics. This proved to be a stimulating venture, and I know that the students who worked with him there will never forget his expertise and his enthusiasm.

Before continuing with the main thesis of this paper, I should like to add to the anecdotal evidence about children and language learning.

101

Although I am solely responsible for any errors made when I mention Hebrew, let me say that I have referred to a fine book on the language recently written by a colleague of Professor Bronstein's and mine (Berman, 1978).

When we arrived in Israel a few days before September 1, the day when all schools in the country begin, my three boys, James, Geoff, and John, were aged James 14.11, Geoff 12.9, and John 10.1. It was probably that these are the only children in the language acquisition literature who had four parents, mother and father and step-parents, each of whom held a Ph.D in linguistics. The possible relevance of this fact may emerge later. Currently, James is doing doctoral work in entomology, Geoff is working for a master's degree in herpetology, and John for a bachelor's in botany. All three, however, have an interest in language and one kind of facility or another in learning foreign languages, depending on the aspect of language under discussion.

Since John, the youngest of the boys, had just turned ten while we were en route, consideration was taken of what Penfield and others had written and the decision was made to enter him in the local Israeli school, *kita he,* the equivalent of the fifth grade. I had much earlier been concerned because a somewhat traumatic episode had seemed to put a stop to John's burgeoning language acquisition in his eleventh and twelfth months. He said nothing for almost two years thereafter. At the age of three, he began to speak normally and had so continued from that year on. So it seemed that he might be still young enough to pick up Hebrew with ease and thereby have a mind-enhancing experience of coming to know a non-Indo-European language fluently. "The best-laid plans of mice and [wo]men. . . ." A young mother in the neighborhood was hired to tutor John in Hebrew since all the classes were to be taught in that language, naturally, in the local public school. This meant that John was exposed to systematic training two afternoons a week, an hour each time, one a one-to-one basis. Emphasis was put on grapho-phonemics and phonetics since, as everybody knows, Hebrew is printed in a different alphabet which reads from right to left. As everybody may not know, however, the handwritten form must also be learned simultaneously because the teacher uses that quite different set of graphemes on the blackboard and the pupils do their work in it while everybody uses the printed form for reading, aloud or to one's self. John's tutor also used short dialogues for practice in reading and pronunciation.

John was, of course, exposed to spoken Hebrew in the classroom, on the playground, and by neighbor children. He was expected to pick up both forms of written Hebrew in class. Naturally, if naively, he was also expected to learn mathematics, Bible studies, Israeli history, and science.

A month passed, and while John was obviously making a real effort to do everything that was expected of him, only the paltry two hours of tutoring seemed to be paying off. I knew no Hebrew to speak of and lectured at the University in English. However, I went to an *ulpanit,* short-term language school, two hours a night, two nights a week, and also practiced a little with the Bronsteins. Consequently, I was able to help my youngest son to some degree.

Still, as you might assume, John's frustration was so great since he was learning no subject matter and would not or could not (there is no way of knowing) utter a word of Hebrew, some change was necessary. Consequently, he joined James and Geoff at the American school and "learned no more Hebrew."

During the month of John's distress, the only reliefs being the short school day, a very kindly *ozeret* (a maid who came every afternoon), and a growing friendship with an English-speaking friend of mine whose children spoke only Hebrew—during these days, how were James and Geoff faring?

Geoff was almost 13, and his chief contact with Hebrew was a class three times a week in the American school, K–12, that serviced the children of diplomats, military personnel, and the like, in the Tel Aviv area. Articulate and outgoing, he would often hitchhike and in the process expand his spoken Hebrew vocabulary. He did this with some success, which he reinforced by playing chess with a Hebrew-speaking Israeli neighbor. He also enjoyed trying to communicate with the *ozeret* but, on the whole, many more hours were spent by Geoff in an English-speaking environment. James, nearly 15, also attended a Hebrew class three times a week at the American School and, at first, the same minimal exposure or less from the *ozeret* and the next-door chess-playing neighbor. After less than two months in Israel, he became, one might say, enamoured of the language and asked if he could take lessons. Delighted to comply, I found him a teacher, a sabra who had just returned from several years with her husband at Stanford University where she had run Hebrew classes for Hillel students who had been of very high caliber. James could hardly wait to got to his two-hour lessons twice weekly and do homework in addition to that from his American school. He very soon outstripped the rest of us in ease of reading and writing, in learning grammatical paradigms and in constructing written sentences. His teacher, whose standards were high indeed, told me several months later that James astonished her by the speed and care with which he was learning. Once he learned that there was a 'Yemenite accent,' he even affected the archaic pronunciation of *alef* as [?], a glottal stop, and *ayin* [ʕ], a pharyngeal fricative, long since gone from other forms of modern Israeli Hebrew. He delighted in giving me help with my homework, liberally sprinkled with criticism. Why could I, a linguist, (even if I was over 30) not always pronounce a *resh* as a uvular fricative instead of using an American *r*? Why could I not distinguish between *xet* and *xaf*? I had a legitimate answer for the latter question (i.e., none but the most pedantic Israeli would make the distinction either), but this did not dispel my young scholar's disapproval.

In short, James became a dedicated student of what is traditionally taught in high school and elementary university classrooms. The fact remained that he could not carry on a fluent spontaneous converstation. Converse he could, however, haltingly, with fine phonological and grammatical accuracy.

By the end of the academic year, when I felt myself on the verge of a breakthrough in conversational Hebrew and had a smattering of grammatical knowledge, where did James, Geoff, and John stand at 15, 13, and 10? James had considerable understanding of structure and some conversational ability; Geoff had adequate if limited conversational knowledge; John, so far as we knew, could only be said to have picked up a few words, seldom used.

Many questions came to mind, some of them partly answered in a recent book on SLA (second language acquisition) (Hatch, 1978). Is second-language learning like the acquisition of a first language? Perhaps the answer is yes under certain conditions, very carefully stated; no, under other conditions. Is SLA monolithic? No is my deeply felt answer when it comes to children of school age and when reading and writing and conscious grammatical knowledge enter the picture. As has been obvious to linguists who have also engaged in language teaching, there is a cluster of language abilities. Some students excel in perception of the foreign language and may also be excellent producers of native-sounding forms, or they may on the other hand get very high marks on dictations but have strong first-language interference when they speak. Other students may or may not be good at the above-mentioned activities but may far excel the first group in their understanding of morphological and syntactic structures. These and other abilities depend, as always, on a combination of genetic factors and the conditioning factors to which a student is exposed in and out of the classroom.

What does a follow-up of James, Geoff, and John reveal? Let me again present some data and leave it to my colleagues whose expertise in SLA is more specialized than my own.

Ilse Lehiste reminds us in the collection of essays in honor of Wise (Bronstein, 1970) that linguists (Jakobson, Fant, and Halle in this case) have always felt that "We speak in order to be heard in order to be understood."

This observation seems most appropriate to Geoff who had perhaps the most "normal" exposure to a foreign language of American school children abroad. I had occasion almost ten years to the day from when we left Israel to discuss his experience with Hebrew since we had left. One period of casual exposure took place when I hired a young engaged Israeli couple to come to our house to help me expand my knowledge of the language, during our first year back from Israel. Once a week Geoff would engage in casual brief conversation with them before we adults went on with our business. James was living with his father in another state during this period, and John was generally off playing with his friends. Geoff's only other experience with spoken or written Hebrew came during a 48-hour stay in our home (six years after we had left Israel) by a young man who was the son of an Israeli friend. Geoff was interested enough to try to engage in conversation in Hebrew for several hours at various times throughout the two days. He reported that a great many words came back to him and also a quite surprising facility that improved in even so short a time. Ten years after we had left, and three years after Geoff had left home, he returned for a visit and I showed him some high-frequency words in a hand-written Israeli letter and was truly surprised to discover that he could still sound out the letters. I also quizzed him on speaking, asking if he remembered how to say some words and even sentences that used to occur frequently in our Israeli environment. He reported that he felt that a return visit to Israel of at least three months would enable him to speak and understand Hebrew quite well even if he did not succeed in reading or writing with ease.

James also came back for a visit ten years after our return, he who had fallen in love with the language; the only thing that reminded him of it in the interval was some tutoring in Egyptian Arabic. When asked to try at least to decipher some material in both Hebrew graphemic systems, he demurred on the grounds that he had neither read nor spoken the language for a decade. I chose material from a printed grammar and from grammatical notes of my own and after a brief period of inspection James announced that he could still read off things like forms in a paradigm and gave evidence of being able to do so quickly and accurately. We both doubted, however, that he could carry on a conversation without extended total immersion in Hebrew. If the reader recalls the nature of James's training, these *ad hoc* test results will not surprise him or her.

My readers may be quite surprised by John's experience, my youngest son, who by all rights should have forgotten everything since he had seemed to learn nothing. How often we underestimate incidental learning.

My husband and I, seven years after the boys and I had returned from Israel, were attending a small weekly two-hour night class taught by a sabra graduate student in biology at our local university. John, by now a freshman in that same university, decided to go with us. The teacher had had some experience in teaching in an *ulpan* and used the mim-men dialogue methods along with questions based on dialogues. We imitated her as a group at first and then singly. To my complete astonishment, John showed that he had completely and accurately internalized Hebrew phonology. The teacher said, "Your mother warned me that you hadn't learned any Hebrew, and here you are speaking like a sabra." He had evidently done some internalizing of syntactic rules as well since he handled the questions well. Dismayed by my lack of faith in what linguists have long been saying about SLA by younger children, distressed by my lack of faith in John's ability, and delighted to be wrong on both counts, I leave you with the puzzle. Why did John appear at the age of nine to have learned no Hebrew? Is there any connection with his having started, then stopped speaking English until he was three years old? How many unanswered questions do we have yet to face in the field of children's language learning?

REFERENCES

Berman, R. A. *Modern Hebrew structure.* Tel Aviv: University Publishing Projects, 1978.
Bronstein, A. J. *The pronunciation of American English.* Englewood Cliffs, N.J.: Prentice-Hall, 1960.
Bronstein, A. J., Shaver, C. L. & Stevens, C. *Essays in honor of Claude M. Wise.* Hannibal, Mo.: Standard Printing Company, 1970.
Hatch, E. M. (Ed.). *Second language acquisition: A book of readings.* Rowley, Mass.: Newbury House, 1978.

13

Generative Generative Phonology

D. Terence Langendoen

Linguistics Program, Brooklyn College, and the Ph.D. Program in Linguistics, the Graduate School of the City University of New York

The syntactic component σ of a generative grammar G can be thought of as a 4-tuple (N, T, A, R), in which N is the nonterminal vocabulary, T the terminal vocabulary, A the axioms, and R the syntactic rules. It is often assumed that the elements of T, the words of a grammar, can simply be listed in a word-dictionary. However, they cannot, as elementary considerations of the productivity of the rules of word formation immediately show us. To take a very simple example, consider the process whereby the prefix *re* can be joined to English verbs, to form new, morphologically complex, verbs. That process is not only productive, in the sense that given practically any verb in English, a new verb can be formed from it by prefixing *re;* it is also recursive, since *re* can be added freely to verbs already containing the prefix. Thus we have, in English, such infinite sets of words as {analyze, reanalyze, re-reanalyze, re-re-reanalyze, . . .}. Consequently, the elements of T cannot simply be listed. They must be generated.[1]

To generate the elements of T, we set up a morphological component μ of G of the form (N', T', A', R'). The nonterminal vocabulary N' of μ consists of morphological categories, one member of which, we may suppose, is W (for 'word'), which is the axiom A' of μ. The members of R' are the morphotactic rules and the elements of T' are the morphemes. Again, it is often assumed that the elements of T' can simply be listed in a morpheme-dictionary, and again that assumption is false. As Halle (1962)

107

shows, there are more morphemes in a language than those which are associated with particular meanings or which combine with other morphemes to form meaningful words. In any language there are also meaningless morphemes that combine with no other morphemes to form meaningful words. In order for a string of phonemes to qualify as a morpheme in a language, it simply has to obey the phonotactic conditions of that language.[2] Now consider a language, such as English, which imposes no limitation on the length of its morphemes (where the length of a morpheme is the number of phonemes in it).[3] In such a language, there is no longest morpheme, and the set of morphemes is infinite.

Thus the morphemes of a language, like its words, must be generated. To generate them, we postulate a phonological component ϕ of G of the form (N'', T'', A'', R'').[4] The nonterminal vocabulary N'' of ϕ consists of the categories of phonology, including the category M (for 'morpheme') that we take to be the axiom A'' of ϕ. The rules R'' are the phonotactic rules and the members of T'' are the phonemes.[5] In the remaining discussion, we examine some aspects of the phonological component of English.

Besides M, we consider the membership of N'' to include S^3 (syllable cluster), S^2(syllable), S^1 (sonant cluster), S^0 (sonant), C^1 (consonant cluster,), and C^0 (consonant). As the notation for representing these categories suggests, the internal structure of a morpheme is a 'projection' of the segmental categories S and C.[6] A consonant cluster is headed by a consonant, which may be flanked by other consonant clusters (typically, but not necessarily, made up of single consonants). A sonant cluster is headed by a sonant, which may be flanked by other sonant clusters (again typically, but not necessarily, made up of single sonants).

A syllable, in turn, is headed by a sonant cluster, which may be flanked by consonant clusters. The fact that a syllable is headed by a sonant cluster, which in turn is headed by a sonant, explains our use of S^2 as the category symbol for syllables. Syllables are projections, ultimately, of sonants. Finally, a syllable cluster is headed by a syllable, which may be flanked by syllable clusters.

A morpheme in English consists either of a consonant cluster, or of a syllable preceded by at most one syllable cluster and followed by at most two syllable clusters. Accordingly, the categorical rule schemata of the phonological component of English are those in (1):

(1) a. $M \rightarrow \{C^1, (S^3) \ S^2 \ (S^3 \ (S^3))\}$
    b. $S^3 \rightarrow (S^3) \ S^2 \ (S^3)$
    c. $S^2 \rightarrow (C^1) \ S^1 \ (C^1)$
    d. $S^1 \rightarrow (S^1) \ S^0 \ (S^1)$
    e. $C^1 \rightarrow (C^1) \ C^0 \ (C^1)$

Like the categorical rules of the base subcomponent of the syntactic component, the categorical rules of the phonological component are context-
free, and they associate with the terminal strings that are eventually generated structural descriptions in the form of tree diagrams or labeled
bracketings. We illustrate this fact by diagramming in (2), (3), and (4) the
structures that the schemata in (1) associate with various English morphemes, assuming the appropriate application of rules for inserting particular phonemes.[7]

(2) a. *th*, /θ/

$$
\begin{array}{c}
M \\
| \\
C^1 \\
| \\
C^0 \\
| \\
\Theta
\end{array}
$$

b. *st*, /st/

$$
\begin{array}{c}
M \\
| \\
C^1 \\
\diagup\;\diagdown \\
C^1 \qquad C^0 \\
| \qquad\quad | \\
C^0 \qquad\quad | \\
| \qquad\quad | \\
s \qquad\quad t
\end{array}
$$

(3) a. *ness*, /nəs/

$$
\begin{array}{c}
M \\
| \\
S^2 \\
\diagup | \diagdown \\
C^1 \quad S^1 \quad C^1 \\
| \quad\; | \quad\; | \\
C^0 \quad S^0 \quad C^0 \\
| \quad\; | \quad\; | \\
n \quad\; \partial \quad\; s
\end{array}
$$

b. *trans, /tránz/*

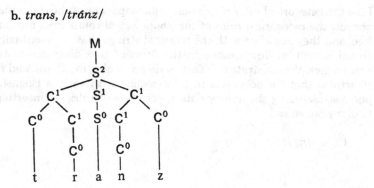

c. *curl, /kérl/*

(figure)

d. *yelp, /yélp/*

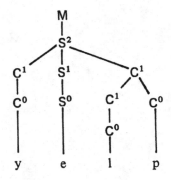

e. *feud, /fyūd/*

f. *yule, /yū́l/*

g. *scowl, /skául/*

h. *scour,* /skáwr/

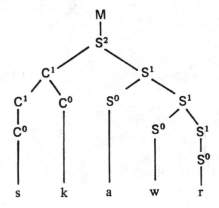

i. *point,* /póynt/

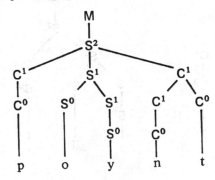

j. *swift,* /swíft/

k. *strict, /stríkt/*

l. *next, /nékst/*

(4) a. *contra, /kóntrə/*

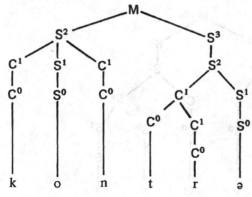

b. *able*, /əbíl/

c. *able*, /ábil/

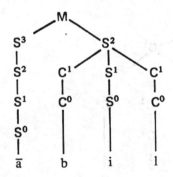

d. *Arthur*, /árθur/

e. *money*, /mónē/

f. *canoe*, /kənú/

g. *modest*, /módəst/

h. *robust, /ròbʌ́st/ or /róbʌ̀st/*

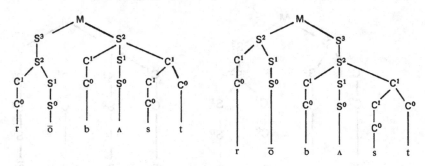

i. *hurricane, /hérəkè̄n/*

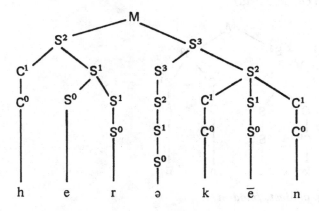

j. *pelican, /péləkən/*

k. schenanigan, /šənánəgən/

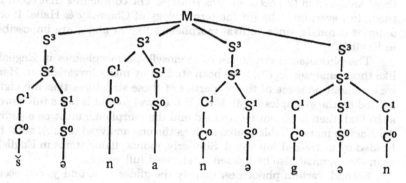

l. salamander, /sáləmàndr/

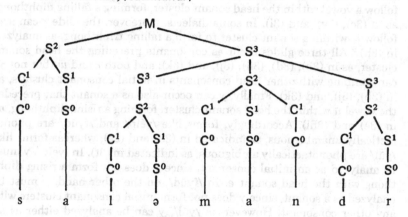

m. abracadabra, ábrəkədábrə/

In English, morphemes consisting of a consonant cluster only, like those analyzed in (2), occur only as suffixes. The consonants that occur in them, however, must be (in the terminology of Chomsky & Halle, 1968) anterior coronals, since suffixal morphemes like *m* and *g* are impossible in English.[8]

The phonotactic structures of monosyllabic morphemes in English, like those analyzed in (3), have been studied by many investigators. Here we comment on some of the properties of those structures that are highlighted by the examples in (3). First, if the head sonant is not a full vowel, as in (3a), then it cannot be stressed and the morpheme must be a suffix. An English monosyllabic prefix, such as the one analyzed in (3b), must be headed by a stressed full vowel. Similarly, monosyllabic stems in English, as in (3c-1), must also be headed by stressed full vowels.

Second, certain phonemes, namely the glides *r, w,* and *y,* can occur either as sonants or as consonants. All three glides occur as sonants that follow a vowel within the head sonant cluster, forming a falling diphthong, as in (3c), (3g), and (3i). In some dialects, moreover, the glide *r* can also follow *w* within a sonant cluster to form a falling triphthong, as analyzed in (3h).[9] All three glides occur as consonants preceding the head sonant cluster, as in (3b), (3d), (3fi), (3j), and (3k); and both *r* and *w* (but not *y*) can combine with other head consonants in initial consonant clusters, as in (3b), (3j), and (3k). Finally, *y* can occur also as a sonant that precedes the vowel *u* within the head sonant cluster, forming a rising diphthong, as in (3e) and (3fii). Accordingly, forms like /yélp/ and /fyūd/ are phonotactically unambiguous, as indicated in (3d) and (3e); whereas forms like /yūl/ are phonotactically ambiguous, as indicated in (3f). In /yelp/, *y* must be analyzed as an initial consonant, since *y* does not form a rising diphthong with the head sonant *e*. In /fyūd/, on the other hand, *y* must be analyzed as a sonant, since *y* does not form initial consonant clusters with any other consonant. However, in /yūl/, *y* can be analyzed either as an initial consonant or as the first member of a rising diphthong.[10]

Third, the internal structure of the syllable establishes the order in which the segments are inserted. Consonants are inserted before sonants, and within clusters modifiers are inserted before heads. In this respect, segmental insertion in phonology is like lexical insertion in syntax.[11] As a result, the insertion of modifying consonants is the least constrained contextually, whereas the insertion of the head sonant is the most constrained. For example, the insertion of *s* in (3g), /skáwl/; (3h), /skáwr/; (3k), /stríkt/; and (3l), /nékst/ is unconstrained by the choice of the neighboring phonemes and is constrained only by its positions within the syllable structure (e.g., as the initial modifier of an initial consonant cluster, or as the initial modifier of the final modifier of a final consonant cluster). Indeed, in those positions, *s* is the only phoneme that can be chosen. On the other hand, the insertion of s in (3j), /swíft/, is constrained by the choice of the following w (any of the phonemes θ, s, t, d, k, or g can occur as the head consonant of an initial consonant cluster when modified by a following w); if r had been chosen instead of w, then s could not have been inserted as the head of the initial consonant cluster (whereas any of the phonemes p, b, f, θ, t, d, $š$, k, or g could have been inserted).

Fourth, within a cluster, the choice of head is based on a principle of relative sonority, which states that the head of a sonant cluster may not be less sonorant than any of its modifiers (including consonantal modifiers) and that the head of a consonant cluster may not be more sonorant than any of its modifiers.[12] In case a cluster is made up of two segments of equal sonority, then the decision as to which is head is based on internal criteria. For example, in the cluster kt that appears in (3k), /stríkt/, we select k as the head and t as the modifier, since this choice makes for a simpler set of phonotactic rules. On the other hand, in the cluster ft that appears in (3j), /swíft/, t is the head and f is the modifier, since f is more sonorant than t. In most cases, including the one just mentioned, the decision to classify segments as heads or modifiers on the basis of the relative sonority principle results in the choice of the simplest set of phonotactic rules. One exception that I am aware of is the classification of s as the head of initial sm and sn clusters in such morphemes as *smear*, /smêr/, and *sneer*, /snêr/. If we were to take m and n as head and s as modifier in these clusters, contrary to the relative sonority principle, then we would have to mention the phonemes m and n in only one phonotactic rule schema having to do with initial consonant clusters (namely, one in which they are inserted as heads following a modifying s); whereas if we treat s as head and m and n as modifiers, in conformity with the relative sonority principle, then the phonemes m and n must be mentioned in two such phonotactic rule schemata (one in which they are inserted in modifier position following a head consonant, along with l, r, w, and y; and another in which s alone is inserted as head before m, n, l, or w). Despite the added complexity that results from treating s as head consonant in initial sm and sn clusters in English, I prefer that treatment to one which requires abandoning the relative sonority principle in this case.[13]

We turn finally to the polysyllabic morphemes that are analyzed in (4). Perhaps the most salient property of these analyses is their resemblance to the metrical trees of Liberman and Prince (1977). In particular, the symbol S^2 that occurs in them corresponds to Liberman and Prince's S (for 'strong'), and the symbol S^3 corresponds to their W (for 'weak'). This correspondence is not accidental, since we can consider the S^2s that are introduced by rule schemata (1a) and (1b) to be nuclear, that is metrically 'strong'; and the S^3s that are introduced by those schemata to be peripheral, that is metrically 'weak'.[14] However, unlike Liberman and

Prince's trees, the trees in (3) and (4) include ternary- and even quater-
nary-branching structures. The reason for our departure from Liberman
and Prince's strict adherence to binary branching is our adoption of the
principles of the $\overline{X}$-theory of constituent structure for phonotactics,
which does not grant any special significance to binary branching.

As in Liberman and Prince's analysis of stress assignment, the relative stress prominence of a syllable in a polysyllabic morpheme can be
determined by its position in the tree structure for that morpheme.[15]. Primary stress falls on the head sonant of that syllable in the morpheme that
is immediately dominated by M. In general, the relative prominence of a
syllable in a polysyllabic morpheme is inversely related to the number of
occurrences of S^3 that intervene between it and M in the tree diagram for
that morpheme. Thus, in the bisyllabic morphemes in (4a–h), the stressed
syllable is immediately dominated by M, while the unstressed syllable is
immediately dominated by S^3, which is immediately dominated by M. A
similar situation holds for the trisyllabic morpheme (4j), /péləken/, and
for the tetrasyllabic morpheme (4k), /šənánəgən/. In the morphemes (4i),
/hérəkèn/, and (4l), /sáləmàndr/, the primary-stressed syllables are immediately dominated by M, the secondary-stressed syllables are separated
from M by exactly one occurrence of S^3, and the unstressed syllables are
separated from M by exactly two occurrences of S^3. Finally, in (4m), /àbrə
kədábrə/, we note that the stress pattern accords with the analysis,
together with the principle that syllables that are not headed by full vowels are always unstressed (cf. note 14).[16]

As the two analyses of (4e), /mónē/, illustrate, there is a certain
amount of ambiguity that arises because of different possibilities of syllable division within morphemes. This type of ambiguity is like the phonotactic ambiguity of morphemes like (3f), /yūl/, noted above, and requires
no further discussion.[17] Another type of ambiguity can arise depending on
which syllable is made head and which are made modifiers. As the analysis
of (4h) illustrates, this ambiguity is reflected in the stress patterns that
are assigned to the morphemes.

While the choice between /ròbʌ́st/ and /róbʌ̀st/ is, in my dialect, completely free, not all such choices are as free as that. For example, the
choice between *Tennessee*, /tènəsé/, and *Tennessee*, /ténəsè/, depends on
whether the word consisting of this morpheme occurs as head of its phrase
or not, as illustrated in (5).

(5) a. Arthur visited Tennessee (/tènəsé/).
    b. Arthur visited the Tennessee (/ténəsē/) Valley.
    c. Arthur visited Tennessee's (/ténəsè/) capital.

This patterning clearly shows that the two variants, /tènəsé/ and /ténəsè/, are indeed two different morphemes, and that a choice between them may be made on syntactic grounds.

I conclude this discussion of the analysis of English phonotactics by noting that the order in which segments are inserted into polysyllabic morphemes follows the same logic as the order of insertion of segments into monosyllabic or nonsyllabic morphemes. First the segments of the most deeply embedded modifying syllables are inserted (those with the least degree of stress), followed by the insertion of the segments of the next most deeply embedded modifying syllables, and so forth, until the head syllable is reached. Thus the insertion of the segments appearing in stressed syllables can be made dependent on the occurrence of segments in unstressed syllables, but not conversely. There is abundant evidence for the correctness of this assumption, but the discussion and analysis of that evidence must await another occasion.

NOTES

1. Not only the morphological structures, but also the semantic structures, of the elements of T must be generated. How this is done is beyond the scope of this essay; for discussion, see Langendoen (1982).

2. I assume that each morpheme has an underlying phonemic shape, which may be altered by application of morphophonemic rules, the discussion of which is also beyond the scope of this essay.

3. I take the claim that there is no phonotactic limitation on the length of English morphemes as not requiring detailed justification. There is, to be sure, a longest meaningful morpheme in English, but its length (whatever it is) is not a consequence of a phonotactic limitation.

4. Once again, we omit consideration of the semantic properties of morphemes. We may assume that one of the tasks of the semantic component of a generative grammar is to generate a set of semantic structures, a finite subset of which is associated with morphemes of the language by an arbitrary pairing.

5. The phonological component ϕ of G is thus very much like the 'phonological grammar' of Householder (1959).

6. The notation is that of the $\overline{X}$-theory of phrase-structure grammar, proposed originally by Chomsky (1970) and developed in detail by Jackendoff (1977).

7. We give the morphemes first as they are spelled in English and second in their underlying phonemic forms. Phonemic representations, when not given in tree-diagrammatic forms, are enclosed within solidi.

8. Note that this is a restriction of the underlying phonological forms of morphemes, not on their surface forms. The contracted form m (from am, /ám/) is not coronal, and the contracted form r (from are, /ár/) is not anterior.

9. In those dialects in which forms like *scour* are disyllabic, the analysis is as in (i):

    (i) scour, /skáwr/

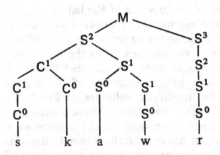

10. The problem then arises of accounting for the fact that the morphemes represented by /yūl/ are associated with the same semantic structure. I assume that this is not a problem that requires a solution within English grammar but that it is a problem for linguistic theory. Essentially what is needed is a principle from which it follows that identical sequences of phonemes in a language receive the same pairings with semantic structures. We may presume that this principle (whatever it is) establishes the general range within which so-called 'free-variants' can vary both phonemically and phonotactically.

11. In syntax, nouns (like consonants) are inserted before verbs (like sonants), and modifiers are inserted before heads; cf. Chomsky (1965, chapter 2).

12. I am indebted to Janet Fodor for this insight. I assume that full vowels are more sonorant than reduced vowels, which are more sonorant than glides, which are more sonorant than liquids, which are more sonorant than nasals, which are more sonorant than aspirates, which are more sonorant than fricatives, which are more sonorant than affricates, which are more sonorant than stops.

13. One important consequence of the relative sonority principle is that initial consonant clusters like *str* in (3j), /stríkt/, and final consonant clusters like *mps* in *glimpse* /glímps/, must be analyzed as ternary-branching. Thus we predict that the insertion of certain head consonants may depend on the prior choice of both the modifier that precedes it and the modifier that follows it. As a case in point, consider the rule schema that inserts the head *t* in initial consonant clusters. If no modifier precedes, then either *r* or *w* can follow, as in (3b) /tránz/, and *twist,* /twíst/. However, if *s* precedes, then only *r* can follow (/stwíkt/ is not a possible English morpheme; at best it will be heard as a misrendering of /stríkt/).

14. Except that an S^2 that is headed by a sonant that is not a full vowel is necessarily metrically weak, as in (3a), /nɔs/.

15. Liberman and Prince's trees actually analyze words, not morphemes. It is

beyond the scope of this essay to consider the analysis of English word stress and metrical structure; see Langendoen (1982) for discussion.

16. We also assume the converse principle of Chomsky and Halle (1968), namely, that syllables headed by full vowels that fall in an unstressed position *in a word* are reduced. Thus, for example, the vowel *i* that heads the second syllable in (4b), /ábil/ is reduced in the word *able*. On the other hand, in the word *ability*, it is the vowel *a* of the first syllable that reduces thanks to this principle.

17. Except to point out that certain investigators prefer to think that bisyllabic morphemes like (4e), /móne̅/, have a single phonotactic analysis, in which the intervocalic consonant simultaneously belongs to both syllables (see Kahn, 1976, for a detailed justification of this position). Obviously, such a treatment cannot be accommodated within the framework presented here, since we assume that the phonotactic structures of morphemes are generated by constituent-structure grammars. In defense of our position, it may be noted that it treats the analysis of morphemes like /móne̅/ on a par with the analysis of (3f), /yúl/, and (4h), /ròbʌst/ or /róbʌst/, namely as falling within the scope of a principle of free variation (see note 10). This is a generalization that is not expressible in frameworks like Kahn's.

REFERENCES

Chomsky, N. *Aspects of the theory of syntax.* Cambridge, Mass.: M.I.T. Press, 1965.

Chomsky, N. Remarks on nominalization. In R. A. Jacobs and P. S. Rosenbaum (Eds.), *Readings in English transformational grammar.* Waltham, Mass.: Ginn, 1970.

Chomsky, N. & Halle, M. *The sound pattern of English.* New York: Harper and Row, 1968.

Halle, M. Phonology in generative grammar. *Word,* 1962, *18,* 54–72.

Householder, F. W. On linguistic primes. *Word,* 1959, *15,* 231–239.

Jackendoff, R. S. *X̄-syntax: A study of phrase structure. (Linguistic inquiry monographs,* 2). Cambridge, Mass: M.I.T. Press, 1977.

Kahn, D. *Syllable-based generalizations in English phonology.* Bloomington, Ind.: Indiana University Linguistics Club, 1976.

Langendoen, D. T. The grammatical analysis of texts, In S. Allén (Ed.), *Text processing: Proceedings of Nobel Symposium 51.* Stockholm, Sweden: Almquist-Wiksell, 1982.

Liberman, M.,& Prince, A. On stress and linguistic rhythm. *Linguistic Inquiry,* 1977, *8,* 249–336.

14

A Note on Replies

Samuel R. Levin

Department of English, Hunter College, and the Graduate School of the City University of New York

In G. Lakoff (1972) it is proposed that the logical form of sentences should include representation of illocutionary forces, *viz.*,

$$\left\{ \begin{array}{l} \text{order} \\ \text{ask} \\ \text{state} \end{array} \right\} (x, y, S_1),$$

where the performative verbs are predicates, x and y are arguments to be carried by grammatical rules into I and *you* respectively, and S_1 is the direct object of the higher verb and, when expanded, will represent the propositional content of S. In discussing the statement alternative Lakoff notes that if someone says

  (1) I state that I am innocent.

and the reply is given

  (R1a) That's false.

what is being denied is that the speaker is innocent, not that he has made the statement. Thus the analysis makes clear the sense in which (R1a) applies only to the propositional content of (1).

The point may be made, however, that it is possible to respond also to the illocutionary force of (1). Thus another possible reply to (1) would be

(R1b) You're lying.

(R1b) is a judgment on the act that the speaker has performed in uttering (1). It is thus a response to the illocutionary force (cf. Austin, 1962). It is not, however, a denial that in uttering (1) the speaker has made a statement (that would be to deny in the face of facts). Instead, (R1b) responds to the presupposition that a speaker in making a statement believes what he asserts to be true. It charges the speaker with not having satisfied that presupposition.

That statements have the bipartite composition expressed by Lakoff's analysis can be shown in another way also. Two other possible replies to (1) are

(R1c) I don't believe it.

(R1d) I don't believe you.

(R1c) asserts that the respondent does not believe the propositional content of (1) to be true, while not commenting on the speaker's presupposed propositional attitude, whereas (R1d) asserts that he has reason to suspect that the speaker knows the proposition to be false and hence does not believe it to be true. Here again, as in (R1b), the presupposition is not satisfied.

In general, where statements are concerned it is possible to question the truth of the proposition or impugn the veracity of the speaker.

Replies like (R1b) and (R1d) are, of course, possible to utterances not containing explicit performatives. That being the case, it would appear that evidence from replies would offer additional support for the need, discussed in Ross (1970), to posit such performatives in the underlying structure of statements like

(2) Prices slumped.

(3) I am innocent.

REFERENCES

Austin, J. L. *How to do things with words.* New York: Oxford University Press, 1962.

Lakoff, G. Linguistics and natural logic. In D. Davidson & G. Harmon (Eds.), *Semantics of natural language.* Dordrecht, Holland: D. Reidel, 1972.

Ross, J. R. On declarative sentences. In R. A. Jacobs & P. S. Rosenbaum (Eds.), *Readings in English transformational grammar.* Waltham, Mass.: Ginn, 1970.

15

The Phoneme
One of Life's Little Uncertainties

Harry Levitt

The Ph.D. Program in Speech and Hearing Sciences, the Graduate School of the City University of New York

INTRODUCTION

Living with uncertainty is one of the few established facts of modern life (or any life for that matter). Although death and taxes do not fall far behind, the only thing we know for certain is that we know nothing for certain. The earliest philosophers were quick to recognize the nature of uncertainty, but it is only in recent times that attempts have been made to provide a rigorous description of uncertainty, at least for those less uncertain aspects of uncertainty that are amenable to mathematical description. Heisenberg's[1] uncertainty principle represents one such attempt. A related form of uncertainty is that of time–frequency uncertainty in spectrum analysis (Gabor, 1946). The entire field of statistics is concerned with the practical problem of reducing uncertainty in measurement and analysis and the relatively new field of information theory follows from the simple but far-reaching proposition that uncertainty can be quantified, bit by bit, so to speak.

This essay is dedicated to Arthur Bronstein who, as teacher, colleague, and friend, reduced the author's uncertain knowledge of phonetics to a more reasonable level of uncertainty.

And, *so to speak,* what uncertainties lie therein! What is it that we can say about speech, that embroidered tapestry of sounds,[2] mirror of the soul,[3] communicator of emotions and ideas,[4] the core of civilization itself[5]? Very little, in concrete terms. But we can speculate at length on the nature of speech and its many uncertainties. Of great interest is that most uncertain figment of our collective imagination, the phoneme. To paraphrase Socrates, we know nothing but our ignorance—but the first step to knowledge is to know how much we do not know.[6] Would the tools of information theory, the mathematics of measuring uncertainty, be of value in this quest? Let us speculate.

The phoneme, or any other entity which behaves as a phoneme might behave, should a phoneme exist, is veiled in uncertainty. What is a phoneme, and if a suitable definition can be postulated, can all aspects of a phoneme be measured? Since, inevitably, some uncertainty must prevail, can the uncertainties of speech be quantified in a systematic way? We shall attempt to identify the major sources of uncertainty governing the specification and measurement of phonemes.

GETTING THE MEASURE OF UNCERTAINTY

According to the mathematical theory of information (Shannon & Weaver, 1949), the information conveyed by an event (observation) is inversely related to the probability of the event. For an event having one of two equally likely outcomes, the information conveyed by the event is one *bit.* In general, the information generated by an event is defined as $I = p \, log_2 \, (1/p)$, where p is the probability of the event and I is measured in bits. In one extreme, when an event is entirely predictable (i.e., p is either 0 or 1) then $p \, log_2 \, (1/p)$ reduces to zero. Thus, no information is generated when the outcome of an event is entirely predictable. At the opposite extreme, a maximum amount of information is generated when the possible outcomes of an event are entirely unpredictable.

The simplest case relates to an event with two possible outcomes, each of which is equally likely. Let the probabilities of the two outcomes be p_1 and p_2, respectively, where $p_1 = p_2 = 0.5$. The information generated by the occurrence of the event is given by

$$I = (p_1 \, log_2 \, 1/p_1) + (p_2 \, log_2 \, 1/p_2),$$

which reduces to

$$I = (0.5 \, log_2 \, 2) + (0.5 \, log_2 \, 2) \; bit.$$

Note that each of the two possible outcomes generates information, as does the non-occurrence of the alternative outcome. This illustrates a basic principle in that information is transmitted not only by what is said but also by what might have been said. Silence is the forgotten phoneme.[7]

The mathematical theory of information provides a numerical basis for quantifying the amount of uncertainty (and hence the amount of information transferred by measuring the reduction in uncertainty produced by a given event). This is an extremely useful method of analysis, but not without its flaws. A major danger associated with the use of information theory is that certain forms of uncertainty are not measurable and the intricate mathematical details of the theory can easily camouflage these other, substantial sources of uncertainty. Thus, for example, a critical factor in assessing the information conveyed by a message is to know the vocabulary of possible symbols making up the message. If the symbol set is wrongly chosen, then the associated measure of information will be in error. If the symbol set is not fully known, then the measure of uncertainty will itself be uncertain.

Herein lies a stark contrast between written and spoken language with the elusive phoneme rising to full glory in a tapestry of interwoven uncertainties. Whereas the symbols of written language are clearly defined and generally accepted, this is not the case with the spoken form of the language. There is no general agreement on the symbol set making up spoken language. Further, speech is not taught, but acquired (with a few notable exceptions involving sensorially handicapped children) and, as a result, the sounds of speech are not fully known. There are also large individual differences and large between-group (dialectal) differences in how speech is spoken.

In order to quantify the uncertainties of speech and, by implication, the information transmitted in a spoken message, it is necessary to know the set of sound symbols (phonemes) that is being used and the probability of occurrence of each of the symbols in the context of spoken messages. In short, in order to quantify uncertainty it is necessary to know the symbol set being used, but this too is uncertain.

Some idea of the range of uncertainty associated with inexact knowledge of the symbol set in speech can be gleaned from several contrasting examples. If the speech signal is considered to be no more than a fluctuating pressure waveform with an effective bandwidth of roughly 20,000 Hz (the upper frequency limit of human hearing) and a dynamic range of about 120dB (this being roughly the intensity ratio of the softest to the loudest speech sounds that can be heard without severe discomfort), then the information that can be transmitted by such a signal is in excess of 800,000 bits/second. According to this view of the speech signal, the symbol set consists of all distinctly different values that the sound pressure waveform can adopt. By distinctly different is meant a value of the wave-form that is measurably different from its neighboring values despite the noise in the system. The size of this symbol set is very large, and the possible rate of information transfer is thus correspondingly large.

The above estimate of information rate assumes that all of the symbols in the set are utilized efficiently, which is not the case. This estimate thus represents an upper bound on possible rates of information transfer. In practice, speech is usually transmitted over communication systems that have much tighter constraints in terms of bandwidth and dynamic range. A typical speech-communication channel might have a bandwidth of 3,500 Hz and a dynamic range of 40dB. Speech heard over such a system would be quite intelligible, although the maximum rate of information transfer is much smaller than before, on the order of 50,000 bits/second.

If speech is viewed as the output of an acoustic system consisting of a pulsive and/or gaussian sound source driving a series of time-varying resonant filters (the conventional acoustic model of speech production), the information rate is closer to 1,000 bits/second. If the basic symbols of spoken communication are considered to be phonemes, then the information transmitted by a totally unpredictable string of English phonemes is roughly 100 bits/second at a normal speaking rate. If the word is considered to be the basic unit of speech, then for a person with a vocabulary of roughly 100,000 words, speaking words randomly at 120 words/minute, the information rate is only about 35 bits/second.

People do not communicate, however, by producing words or phonemes in random sequence, but rather use these symbols to convey thoughts. If thoughts are considered to be the basic unit of communcation, then the information rate is estimated to be under 10 bits/second.

Estimates of the inherent uncertainty of speech and of the reduction in uncertainty produced by a spoken message (i.e., the information transmitted) clearly vary over an enormous range, depending on the symbol set that is used. Our measure of uncertainty is, thus, itself uncertain. It should be apparent from the preceding discussion, however, that there are two distinctly different forms of uncertainty. The first is that form of uncertainty which can be measured and quantified, such as the measure of uncertainty in information theory. The second (not knowing which is the correct symbol set) represents a form of uncertainty which cannot be measured reliably, if at all.

A measure closely related to the information-theoretic measure of uncertainty is that of precision in statistical measurement. By precision is meant the degree to which a particular measurement is repeatable. The standard error of a measurement is a commonly used measure of precision. In principle, precision can be increased (i.e., standard error reduced) by taking the average of several measurements. In order to assess by how much precision is increased by repeated measurement, it is necessary to make some assumptions about the statistical properties of the quantity measured. A common, relatively unrestrictive assumption is that these statistical properties do not change between measurements (e.g., the statistical distribution of the measurements, whatever its form, does not change between measurements).

The two forms of uncertainty, in statistical terms, take on the guise of imprecision and inaccuracy, respectively. If the underlying assumptions are valid, then the precision of measurement can be quantified. If the underlying assumptions are not valid, the resulting estimates will be both imprecise and inaccurate. Unlike precision, degree of accuracy cannot be measured in absolute terms. It is possible, however, to measure relative accuracy; for example, by comparing a new method of measurement with an established, standard method of measurement, but one should bear in mind that the accuracy of the standard method is itself not known in absolute terms.

There is an obvious link between information theory and statistics in that the observer acquires information with each measurement. The amount of information transferred is directly related to the increase in precision obtained per measurement. A more revealing link between the two theories is that the measure of information $[p \log (1/p)]$ and the measure of precision (the standard error) are both measures of the dispersion or spread of an underlying statistical distribution. The standard error of a distribution is an intuitively obvious measure of dispersion; the measure of information, $[p \log (1/p)]$, is also a measure of dispersion with the very special property that it is the measure with the smallest expected value for a given distribution. A very useful practical consequence of this property is that the expected cost of a series of observations can be minimized by making the cost of each observation proportional to $\log (1/p)$.

THE COST OF MEASUREMENT

A fundamental aspect of measurement that is often overlooked is that there is a cost associated with every observation. This cost may take the form of time taken in making the observation, or energy absorbed from the quantity being measured, or other factors which, in some way, either alter the quantity being measured and/or limit the resolution of the measurements. The possible effects of performing experiments on others, on animals, or on the environment in general must also be taken into account. Our concern in this essay is strictly amoral. We are not addressing the question of cost in an ethical sense (an issue of fundamental concern to all concerned researchers), but rather the much narrower issue of the effect of measurement costs on the uncertainties of measurement.

The time taken to make an observation and the requirement that some energy be absorbed as part of the measurement process imposes a fundamental limitation on the resolution of any physical measurement. In the case of atomic particles, it is not possible to know both the location and momentum of a particle with unlimited precision (Heisenberg's uncertainty principle). In signal analysis it is not possible to measure both the frequency spectrum and temporal structure of a signal simultaneously with unlimited precision.

Gabor (1946) provided an elegant mathematical formulation of the trading relationship governing time and frequency resolution in spectrum analysis. He showed that as a general principle,

(frequency resolution) x (time resolution) $= K$,
where K is a constant.

There are many ways in which frequency resolution and time resolution can be specified, and analogous definitions of frequency and time resolution must be used in order to interpret the above relationship properly. The constant K depends on the definitions of frequency and time resolution that are used and on the resolving power of the measuring instrument. A common limitation of acoustic measuring instruments, including the ear, is that energy below a certain minimum level is not detected in either the frequency or time domain.

Given the nature of the trading relationship between time and frequency resolution and the limited sensitivity of any measuring instrument, it is possible to quantify the relative uncertainties inherent in the measurement of the spectral and temporal structure of speech signals. As before, this measure of uncertainty is dependent on an underlying model, the mathematical model underlying the concept of spectrum analysis. If this model does not hold, then our measure of time–frequency uncertainty is itself uncertain. Spectrum analysis has been found to be a very useful tool in the analysis of speech signals, and thus the nature of its inherent uncertainties needs to be known.

UNCERTAINTIES IN THE SPECIFICATION AND MEASUREMENT OF PHONEMES

The measurement of phonemes follows directly from their definition. As discussed by Bronstein (1960, p. 22), "a phoneme may be defined as a sound that is significantly different from the other sounds of the language," and in order to identify the phonemes of a language it is necessary "to break the language down into its meaningfully different, or significantly different, sounds." In order for phonemes to be identified, it is necessary for skilled observers to listen and decide which sounds or sound contrasts in a language convey meaning. This is not a simple task. It involves subjective judgment and takes time.

The usual method of identifying the phonemes of a language is for one or more skilled phoneticians to listen to the sounds of the language, to record what was said in phonetic terms, and then to identify which phonetic changes are associated with changes in meaning. The terminology used in specifying the sounds heard is the common invention of phoneticians; the subjective experience of hearing sounds is unique to every human being. Although it is not possible for one human being to experience the experiences of another (even the most empathetic phonetician cannot perceive the phonemes perceived by another), a remarkably high degree of consistency has been observed in the reports of different people on hearing the same utterance.

The degree of consistency in the perception of speech (within groups of people familiar with a given dialect) is so high and so common in our everyday experience that we are sometimes surprised when the usual high degree of consistency does not occur. The belief that someone can "mishear" what was said reflects this attitude. We can never know what another person heard (or misheard), but we do know when someone's reported perception of what was said differs from the majority view. Phonetics is a truly democratic science in that the defined sounds of a language (e.g., the phonemes) reflect the majority view of what is perceived when an utterance is produced.

As in any good democracy, the majority view is not the only view, and there are many differences of opinion on the specifics of how the sounds of a language should be defined. In particular, the concept of the phoneme is under attack, an alternative view being that speech is made up of segments and that each segment, in turn, is made up of a set of distinctive features. These features are binary in form in that they either exist or do not exist for each sound of speech. They are distinctive in that a change in any feature will change the meaning of what is said.

It is interesting to note that the use of binary distinctive features as a basic symbol set for describing speech draws heavily on the ideas of information theory. The binary pair is the simplest symbol set of all. One of the important contributions of information theory was to show how even the most complex symbol set can be reduced efficiently to a minimal number of independent binary symbols. The concept of a minimal set of binary distinctive features for representing a code as complex as speech reflects this type of thinking, but it is curious that distinctive-feature theory does not also take advantage of one of the most useful results of information theory, that of variable-length coding.

The theory of binary distinctive features, both in its original and revised forms (Jacobson, Fant, & Halle, 1951; Fant, 1973), uses an essentially fixed number of binary distinctions for representing each segment of speech, four or five binary features being used to represent the vowels and seven or eight binary features being used to represent the consonants, regardless of their frequency of occurrence. A more efficient binary code is one in which fewer symbols are used for the more common speech segments and a greater number of symbols is used for the less common segments. For maximum efficiency, the number of binary symbols per segment should be proportional to log $1/p$, where p is the frequency of occurrence of the segment.

There is some evidence that the articulatory mechanisms involved in speech production are consistent with the above principle. The speech segments that occur least frequently in spoken English (Denes, 1963) are the longer, compound sounds consisting of two or more components, such as diphthongs and affricates. In contrast, the sounds occurring most frequently (i.e., /ð,i,t,n,s/) cannot be subdivided into shorter, self-contained components. The most frequent sounds of speech typically also involve a common articulatory component (e.g., the vowels produced most frequently are front vowels; the consonants produced most frequently involve the alveolar place of articulation). On the basis of these considerations, it is possible to develop a variable-length binary representation of the sounds of English that is both relatively efficient and directly linked to the articulatory mechanisms of speech production. There is also good reason to believe that speech, being the result of an evolutionary process, is produced efficiently and that the basic symbol set representing speech should also be an efficient code.

Whatever symbol system is used, there will always be room for uncertainty in specifying what was said. A common source of confusion occurs with sounds having a similar articulatory and/or acoustic structure. Does Popeye really say "I yam what I yam," or is it "I am what I yam" or "I yam what I am"? There is little doubt that Popeye means what he says but does he say what he means? The degree of uncertainty between what was intended and what was perceived can be quantified to some extent with the honest cooperation of both the speaker and listener. There is a large body of literature documenting the common perceptual confusions in both the visual (lipreading) and auditory perception of speech (Miller & Nicely, 1956; Erber, 1972), and it is even possible to predict many of these confusions on the basis of the physical characteristics, visual and/or acoustic, of the speech signal (Dubno & Levitt, 1981; Walden, Prosek, & Worthington, 1974).

It is important to bear in mind that the predictions of uncertainty that have been developed are for average frequencies of confusion. It is not possible to predict how a phoneme will be perceived on a single, specific occasion. Note also that the quantification of uncertainty in speech perception has thus far related primarily to context-free conditions, for instance, perceptual confusions involving nonsense syllables and monosyllabic words.

An extremely challenging problem is that of quantifying the uncertainties of speech perception for sounds produced in context. For example, the phoneme sequence /səks/ produced in isolation is likely to be heard as the word *socks* by a draper, as *sax* by a jazz musician and as *sex* by a pair of newlyweds. In order to predict and quantify confusions in speech, it is necessary to know what is on the listener's mind—clearly not an easy task.

The effect of context on speech perception can be measured (and predicted) when context is controlled experimentally. Miller, Heise, and Lichten (1951) in their classic study on the effects of context showed that the intelligibility of speech under difficult listening conditions increased as the predictability of the test words increased. The predictability of test items (e.g., features, phonemes, words, phrases, or sentences) in an intelligibility test can be increased by either reducing the number of response alternatives or by altering the linguistic context in which the test items appear. In either case, increasing the predictability of the test items increases their relative intelligibility. Subsequent to this pivotal study, Green and Birdsall, (1958), Pollack (1964), Boothroyd (1968), and others have shown that the amount of the improvement in intelligibility can be predicted rather well using the mathematics of information theory.

A relatively recent development has been the standardization of tests of speech perception that measure the effect of context directly. In the Speech in Noise (SPIN) test developed by Kalikow, Stevens, and Elliot (1977), the intelligibility of balanced sets of words in both a high and low predictability context is measured. Tests of this type are useful but can only approximate the effects of context in actual communication. Nevertheless, a reduction in uncertainty (*re:* the effects of context) is better than no reduction.

Another source of uncertainty results from the cost of measurement. One of these costs is the time taken in making observations. Spoken language is an evolving process. If the time taken to obtain a complete description of the phonemes of a language is extremely long, then the language itself will have changed by the time the last measurements are obtained. The early measurements will thus relate to one form of the language while the later measurements will relate to another form. It is thus not possible to obtain, by observational means, a description of a language that is complete in every detail while also describing how the language changes with time. Some compromise is necessary between a comprehensive description taken over a long time and a series of less detailed descriptions that map out how the language is changing with time.

Another type of measurement cost is that the quantity being measured may itself be altered as a result of the measurement. For example, in order to measure emerging language in a young child it may be necessary to communicate with the child or instruct the child how to perform a given test. The act of making these measurements may thus affect the child's acquisition of the language (not necessarily in an adverse way). The more detailed or precise the measurements need to be, the greater the number of observations and hence the greater the probability of the measurements' affecting the child's language development. Thus, a detailed, complete description of a child's natural language development cannot be obtained by this means, and a compromise must be drawn between a very detailed series of measurements that is likely to affect a child's language development and a less detailed series of measurements that can track natural language development with little possibility of interference.

Uncertainties in Measuring the Physical Correlates of Phonemes

Measurements of the articulatory or acoustic characteristics of phonemes are subject to the same sources of uncertainty as in the direct subjective measurement of phonemes. Although the observation and specification of physical characteristics typically employ objective methods of measurement, a critical subjective element remains. Someone has to identify the phonemes and which measurements are associated with which phonemes. For example, it is not very difficult to obtain a precise record of either the time waveform or frequency spectrum of the speech signal. It is also possible to identify the essential characteristics of the acoustic system that generated the speech signal. All this can be done quite objectively. What cannot be done without human intervention is the initial judgment as to which of these physical characteristics of the speech signal correspond to a given phoneme.

Measurements of the physical correlates of phonemes, despite their seemingly objective and precise nature, thus rest on an uncertain subjective base. It is important not to lose sight of this inherent source of uncertainty while taking advantage of the extremely high precision of measurement provided by modern instrumental techniques.

A pertinent illustrative example relates to the precision and accuracy with which the speech waveform can be segmented. Presumably objective rules for segmenting the speech signal, either in the form of a time waveform or spectrogram, have been developed (Peterson & Lehiste, 1960; Klatt, 1976), but these in turn are based on earlier subjective judgments of the physical cues signaling phoneme boundaries. The problem is complicated considerably by coarticulatory effects which influence the nature of these physical cues depending on the phonemes involved. This opens up the possibility of confounding the effects of phonetic context and the choice of physical cues for a given phoneme boundary.

Given a well-defined set of rules, the speech waveform can be segmented with considerable precision. The test–retest repeatability of these measured boundaries can be reduced to the order of microseconds using modern digital instrumentation. In contrast, test–retest precision of subjective judgments of phoneme boundaries is several orders of magnitude greater. Estimates of the precision of segmentation by subjective judgment range from just under 10 milliseconds to over 100 milliseconds, depending on the types of phonemes on either side of the boundary, the method of subjective judgment, and who is doing the judging (Osberger, 1979).

The relative accuracy of the two sets of measurements is a separate issue. The subjective judgments of phoneme boundaries have greater face validity, but unfortunately the poor precision of these measurements places severe restrictions on their potential accuracy. One approach is to use the judgments of the most precise and consistent listener in developing a set of objective rules for each boundary condition. This approach has good face validity in that it is the result of direct judgment of the phoneme boundary while at the same time the relative precision of the subjective judgments on which the technique is based is reasonably good. For neither the subjective nor objective method of segmentation can the accuracy of the measurements be determined in absolute terms since the true values of the boundaries are simply not known.

Since the precision of objective measurements is usually so much greater than that of subjective measurement, it is tempting to redefine the quantity to be measured in terms of objectively determined parameters. For example, instead of measuring the acoustic characteristics of phonemes one can measure the acoustic characteristics of speech segments where each segment is defined by specific articulatory behavior, the acoustic correlates of which can be determined with great precision. The quality with which these segments represent the basic components of speech depends on how they are defined and on the purpose of the study. Substituting one type of speech unit for another does not necessarily solve problems, but it can be helpful in reducing the uncertainties of measurement.

Greater precision of measurement does not eliminate uncertainty, but it can be extremely effective in partitioning uncertainty into its various forms, thereby providing a broad perspective on how best to reduce uncertainty by careful experimental design.

Issues of experimental design have not received the attention they deserve in experimental phonetics. Many of the difficulties encountered in measuring the physical correlates of phonemes are a result of the inherent limitations of experimental designs involving random variables and are not peculiar to the concept of a phoneme *per se*. Factors affecting the physical correlates of phonemes include phonetic context, coarticulatory effects, semantic content, between-talker differences, intratalker differences, and method of measurement (e.g., whether these are derived from the time waveform, short-term frequency spectrum of the speech signal, or physiological measurements of articulatory events). Note that for objective measurements, between-observer differences and test–retest variability can be made negligibly small compared to the sources of variance cited above.

It is possible, using a balanced experimental design, to estimate the relative magnitude of each of these effects and the interactions between these effects. It is thus possible to partition out the various factors influencing the measured physical correlates of the phoneme.

A fundamental problem is that while some of the factors affecting these physical correlates are deterministic in their effect (i.e., they produce the same, *fixed* effect on repeated measurement), others have a *random* effect (i.e., repeated measurements result in random variations in the measured effect). An example of a fixed effect is that of coarticulation in a given phonetic context. The physical correlates of the phonemes involved are affected in essentially the same way whenever the same phonetic conditions occur. In contrast, differences between separate productions of the same utterance are typically random in their effect. The specific differences between repeated productions of the same utterance cannot be predicted, although the average magnitude or variance of these differences can be estimated. Between-talker differences may be either fixed or random in their effect depending on whether many measurements are obtained on a few carefully selected talkers (fixed effects) or a few measurements on many, randomly selected talkers (random effects).

The problem of separating fixed from random effects is well known to statisticians. It is a very difficult problem in a multivariate experimental design, particularly if some of the variables have a fixed effect, others have a random effect, and still others have a mixed (fixed and random) effect. This, in essence, is the nature of the problem confronting the experimental phonetician concerned with the measurement of the physical correlates of phonemes. Although various techniques have been developed for dealing with fixed and random effects in experimental design, there are inherent limitations to even the best experimental design.

A fundamental limitation is the inherent uncertainty associated with the random effects. In a well-balanced multivariate experiment consisting solely of fixed effects it is possible not only to estimate each of the effects but also to predict, with a precision comparable to the precision of measurement, all of the measurements in a replication of the experiment. This cannot be done in an experiment involving random effects. At best, the relative contribution of each factor to the overall variance of the measurements can be estimated. An additional disadvantage of the random-effects model is that the statistical tests for separating out the various components of variance are relatively weak and more critically dependent on the underlying assumptions as compared with the comparable statistical tests in a fixed-effects model.

It would be nice if associated with each phoneme there existed an easily measured, invariant set of physical cues. As yet, such has not been found to be the case, although the search for invariance continues. From a statistical viewpoint, these invariant cues correspond to fixed main effects in an experimental design where the phonemes are the primary factor of interest. The relatively large body of acoustic-phonetic data that has been gathered thus far, although not always in the form of balanced experimental designs, shows significant interaction effects which are often so large as to blur the much sought after main effects; that is to say, a single set of invariant cues associated with each phoneme is not always evident.

The problem is compounded by the fact that the interactive components are not necessarily fixed but may be random in their effect. A fixed-effect interaction can at least be estimated, and the physical cues associated with each phoneme in a given context can, in principle, be determined, although the mechanics of doing this may be very difficult. A random-effect interaction, by its nature, does not occur consistently and hence its effect cannot be estimated for each individual occurrence. All that can be estimated is its relative contribution to the overall variance of the measurements.

An added complication is that it is not always easy to distinguish between fixed and random effects, especially if the fixed effect is the sum of many independent components. Although each component may be fixed in its effect, the sum of many independent fixed effects looks very much like a random variable (the law of large numbers). If the structure of the fixed effects is known in advance, then the problem of estimating the components of the fixed effects is simplified considerably.

In essence, random effects generate uncertainty, whereas fixed effects represent a code that can be broken. From this viewpoint, the fundamental problem limiting the automatic recognition of speech by machine is that the acoustic correlates of phonemes (or any other symbols used to describe speech) must be determined experimentally where many of the key variables are random rather than fixed in their effect. A fixed-effect solution can sometimes be forced on an essentially random-effect problem in order to obtain an approximate solution of limited scope. Such a solution may be of practical value although not easily generalized, as appears to be the case with the relative success now being obtained with automatic speech recognition devices for speech constrained to limited vocabularies and cooperative speakers.

Yet another form of uncertainty worthy of note is that involving measurements of finite duration. As noted earlier, there is an inherent trading relationship between frequency resolution and time resolution in the measurement of acoustic or any other signals. The longer the time interval over which the measurement is taken the higher is the frequency resolution, and *vice versa*. Since both the frequency spectrum and temporal structure are of importance in specifying the acoustic characteristics of phonemes, decisions must be made as to which form of uncertainty is less desirable.

In some instances, it is necessary to measure the frequency spectrum with great resolution; for example, when it is of interest to examine the harmonic structure of voiced sounds. In this case a narrowband frequency analyzer is typically used with a concomitant loss in the precision with which temporal changes can be measured. In other instances, it is necessary to measure the temporal characteristics of a phoneme such as, for example, measuring the sequence of events immediately following the release from closure in stop consonants. In this case, a broadband frequency analyzer is typically used with a concomitant loss in the precision with which the fine structure in the frequency spectrum can be measured.

The use of both a narrow (45Hz) and broad (300Hz) bandwidth has become a common practice in the analysis of speech signals. These two bandwidths, however, are often treated with a reverence approaching that of a centuries-old religious tradition. Although these bandwidths are very useful and well suited to the analysis of normal speech, they are not sacrosanct and can, in fact, generate more uncertainty than is necessary. For a voice with abnormally high pitch, as is often the case with a deaf child, the so-called "formant bars" in a standard broadband (300Hz) spectrogram may well be harmonics of the voice's fundamental frequency. A possible way around this dilemma is to use a frequency analyzer in which the analyzing bandwidth is adjusted according to a well-defined adaptive strategy depending on the acoustic characteristics of the signal to be measured. The intelligent use of adaptive techniques can reduce the uncertainties of measurement in many ways, as we shall discuss shortly.

A source of uncertainty analogous to that occurring in short-term frequency analysis is that of tracking time-varying effects (fixed or random) in a lengthy experiment. If a very long experiment is performed, each of the effects, averaged over time, can be estimated with great precision. If a series of short experiments is performed, the time-course of the changes can be mapped out, but the relative precision of each estimate will be reduced. The mathematical similarities between estimating time-dependent effects in an experiment and that of deriving the short-term spectrum of a signal are illustrated in Levitt (1972).

In spectrum analysis, as in statistical analysis, it is important to distinguish between fixed and random effects. The periodic vibrations and well-defined transients produced by the vocal apparatus are the fixed effects of the speech signal. The turbulent airflow produced during a fricative or plosive release is a random component of the speech signal. Not only does the frequency spectrum of the fixed effects of the speech signal provide a meaningful description of both the intensity and phase of these components as a function of frequency, but this information can also be used to produce a consistent reproduction of the time waveform of the speech signal. The frequency spectrum of the random components, on the other hand, provides a useful description of only the relative intensity of the speech signal as a function of frequency, there being no consistent phase information. Since the ear is relatively insensitive to phase, it has become a common practice to discard phase information in analyzing the acoustic characteristics of speech. This is convenient since it avoids the difficulty of distinguishing between fixed and random phase effects.

The process of discarding phase information is one of deliberately increasing uncertainty, the justification being that phase information appears to be of little practical consequence in speech perception. An insidious danger is that phase information may be inadvertently discarded or destroyed in applications where it is of considerable importance. An example of one such error is the use of conventional audio equipment with poor phase characteristics in the measurements of vocal cord vibration where the exact phase of the opening and closing of the glottis is an important consideration.[8]

TOWARD THE EFFICIENT REDUCTION OF UNCERTAINTY

Of the many manifestations of uncertainty that embrace the phoneme, two basic forms stand out: uncertainty that is measurable and that which is not. Uncertainty that can be measured and quantified can also be minimized in systematic ways, and many techniques for reducing the uncertainties of measurement have been developed. The philosophy underlying these techniques and how they may be applied to the reduction of uncertainty in our study of the phoneme, and of speech in general, are of direct interest.

The advent of statistics introduced the concept of measuring uncertainty (e.g., by measuring the statistical spread of random events), and it was not long before systematic methods for increasing precision by improved experimental design were developed. One such development was the idea of a pilot experiment in order to obtain preliminary information on the quantities to be measured. With this preliminary information it is possible to develop a much more efficient design for the experiment to follow.

The concept of the pilot experiment illustrates a fundamental principle of measurement: in order to measure a quantity efficiently, we must know the nature of the quantity being measured. A corollary to this principle is that, in order to measure a quantity with maximum efficiency, we must know exactly that which we wish to measure. This results in a catch-22 situation; if we already know that which we wish to measure, there is no longer a need for the measurement. In practice, we are almost always somewhere between these two extremes, knowing something about the quantity we wish to measure, but not knowing enough so that further measurement is justified.

A practical way of dealing with the above dilemma is to use an adaptive measuring procedure in which every new measurement is made as efficiently as possible using the information available from all previous measurements. The criterion for efficiency depends on the goal of the experiment. For our purposes, an efficient observation is one that maximizes the expected gain in information with each observation. This approach in effect reduces measurable uncertainties in an efficient way.

The extent to which the approach reduces unmeasurable uncertainties cannot be quantified because of the very nature of these uncertainties.

An adaptive experiment need not be a complicated experiment. Simple rules can be developed for efficient adaptive control of experimental variables. A family of procedures of this type are the up-down and transformed up-down techniques (Levitt, 1971) commonly used in psychophysics and now being used increasingly in experimental phonetics. Other, more complex types of experiment may require elaborate calculations before each observation in order to maximize statistical efficiency. The recent development of low-cost, high-powered microcomputers has made the latter type of adaptive experiment an attractive, practical possibility.

The history of the phoneme, its specification, and attempts at its measurement illustrate the general principle described above. In order to measure the phoneme it is necessary to know the phoneme, but the phoneme is itself an uncertain quantity. It is nevertheless apparent that as both our tools of measurement and our understanding of how best to use these tools improve, so the uncertainties surrounding the phoneme are gradually being reduced resulting, in turn, in more efficient methods of measurement.

It is important to distinguish between two basically different factors that have helped increase our understanding of the phoneme. The first is improved instrumentation that has provided measurements of considerable precision of the various physical correlates of the phoneme. The second is our improved understanding of the nature of uncertainty and the application of this knowledge to improved experimental design.

It is hardly necessary to point out the enormous impact of modern instrumentation on phonetics as a science. Prior to the invention of the phonograph, the sounds of speech were measured almost entirely by ear. Since then the instrumental study of speech (and related aspects of hearing) has grown immensely in sophistication such that today these techniques cover almost every known method of measurement. Significantly, each major advance in instrumentation has brought increased understanding of the speech process which, in turn, has resulted in improved designs for the next generation of speech-analyzing instruments.

The impact of improved experimental design on our understanding of the speech process is perhaps more subtle, but no less important. As in the development of improved instrumentation, an evolutionary adaptive mechanism is clearly at work in that, on the average, increasingly more powerful experiments are being designed based on the results obtained in previous experiments.

An illustrative example of this general process is provided by recent research developments in the area of categorical perception. A central experimental concern in this area is whether the slope of the response curve for a just-noticeable difference in sound perception is steeper between sound categories than within categories. Psychophysical response curves are, in essence, cumulative probability distribution functions where the slope of the response curve is inversely related to the dispersion of an underlying probability density function and hence is a measure of uncertainty. The steeper the slope, the smaller is the dispersion of the measurements and the less the uncertainty about the location of the boundary between the two conditions traversed by the response curve.

The problem of comparing the slopes of the two response curves is equivalent to that of comparing the variances of two random variables. As we suggested earlier, statistical tests for discriminating between variances are generally much less powerful than the corresponding statistical tests for discriminating between means (i.e., fixed effects). One of the reasons why this should occur is that any random effect causing the mean of a response curve to fluctuate will not have much effect on the average value of the mean, although the precision of estimation will be reduced. In contrast, the effect of these random fluctuations on the slope of the response curve will always be in the same direction, that of reducing the slope. The effect of these random fluctuations on slope thus cannot simply be averaged out. In fact, the averaging process makes matters worse by steadily reducing the quantity being measured, the slope of the curve, with increasing sample size. Early experiments on categorical perception did not take these statistical considerations into account, with the result that the conclusions drawn were shrouded in uncertainty. Although the issues were not resolved, the experiments provided important information on the nature of the variables involved. Improved experimental designs were soon developed on the basis of this information. These experiments similarly provided new insights, which in turn have resulted in the design of even more powerful experiments. This iterative process has continued and the stage has now been reached where the uncertainties involved (the slopes of the response curves) are themselves being measured by an adaptive process during the course of an experiment (Kewley-Port & Pisoni, 1982).

In short, an increased awareness of the nature of the uncertainties affecting the phoneme and its measurement has led to the development of more effective techniques for reducing these uncertainties. A logical extension of this process is to try to minimize those aspects of uncertainties which can be minimized as a specific experimental objective. An inherent danger to be guarded against is that the process of minimization may place undue weight on additional assumptions of unknown or questionable accuracy. That is, an increase in the theoretical efficiency of the experimental procedures is desirable but not at the expense of the procedures being critically dependent on detailed, unproven assumptions.

As is evident from the above, the development of the mathematical tools for quantifying uncertainty has resulted in significantly more powerful experimental techniques designed to minimize those aspects of uncertainty that can be specified mathematically. This raises an interesting philosophical question with important practical implications. Can our improved understanding of the nature of uncertainty, including the distinction between measurable and unmeasurable uncertainty, be used to reduce further the domain of unmeasurable uncertainty?

The mathematical analysis of uncertainty has resulted in the elucidation of several fundamental principles. One of these is that the precision of measurement is increased (i.e., uncertainty is reduced) by taking the average of several *independent* observations. It seems a reasonable assumption, but one that cannot be proven, that the same principle holds in the domain of unmeasurable uncertainties. Specifically, it is assumed that the accuracy of measurement can be increased by observing a quantity in several independent ways. For example, if each method of measurement is subject to a bias of unknown magnitude that limits the accuracy of measurement, then the process of taking several independent measurements is likely, on the average, to reduce the effect of these biases. The key is to obtain measurements that are truly independent of each other so that there is no reinforcement of common biases.

An appealing speculation is that the above method of reducing the domain of unmeasurable uncertainty could be applied to the phoneme. As noted earlier, the methods of experimental phonetics cover a wide range of measurement techniques. Consequently, attempts have been made to measure the phoneme and its physical correlates in a variety of very different ways, perceptually, acoustically, and physiologically. The fact that the phoneme can be identified and its characteristics measured in so many different ways is strong evidence of its existence. Further, the biases affecting the measurement of the phoneme and its characteristics in these separate domains are, presumably, reasonably independent of each other, thereby providing a global view of the phoneme that, in principle, should be more accurate than that obtainable in any one domain.

One final speculation deals with the presumed reason for the uncertain nature of the phoneme. Although speech is a product of the mind, it is the cumulative result of an evolutionary process. The phoneme (or any other symbol system) is, on the other hand, a deliberate invention of the mind designed for the purpose of specifying speech. The imperfect match between speech and its method of specification is a breeding ground of uncertainty. However, since the process of measurement is an evolution-ary, adaptive process, the degree of mismatch should be gradually reduced as the science of phonetics progresses. A pertinent question is whether or not the concept of the phoneme should itself be refined. The current thrust in phonological theory is toward a description of speech that can be derived directly from the morpheme. The sounds of speech, in this case, are described in terms of phones which, in turn, are made up of fea-tures, the exact structure of which depends on the specific phonological theory that is invoked.

There are a sophistication and depth to modern phonological theories that reflect our current understanding of language itself. On the assump-tion that modern theories are more insightful than older theories, on the average (the underlying theme of any good adaptive process, and it is believed that phonetic science is such a process), then modern phonolog-ical theory should, in principle, provide a more insightful description of speech. But, as we noted earlier, there is a cost to measurement, including both time and effort, and it may be a while before strong experimental evidence is brought to bear showing the measurable superiority of any given theory. The evidence, thus far, indicates that modern phonological theories are able to explain aspects of speech that are beyond the compass of the phoneme. At the same time, these gains are bought at the cost of a greater reliance on theory.

From the vantage point of decreasing uncertainty, the current trend in phonological theory, with its deemphasis, if not total neglect, of the phoneme, brings with it new information as well as new uncertainties of unmeasurable magnitude: that is, the value of the new information depends on the accuracy of the underlying theory which at this stage has yet to be determined. The phoneme, on the other hand, is not dependent on the details of any specific theory but only on the assumption that changes in meaning are conveyed by changes in sound. The phoneme, by definition, describes these changes. The description may be quite complex in many instances, but the concept is simple, elegant, and robust. As Alice learned some time ago, "take care of the sense and the sounds will take care of themselves."[9]

NOTES

1. Also a man of uncertain principles, he was Germany's leading scientist during the Hitler period.
2. "The speech of man is like embroidered tapestries, since like them this too has to be extended in order to display its patterns, but when it is rolled up it conceals and distorts them." (Plutarch: *Lives:* Themistocles, sec. 29)

3. "Speech is a mirror of the soul: as a man speaks, so is he." (Publilius Syrus: Maxim 1073)
4. "With the sense of sight, the idea communicates the emotion, whereas with sound, the emotion communicates the idea, which is more direct and therefore more powerful." (Alfred North Whitehead: *Dialogues,* p. 231)
5. "Speech is civilization itself. The word, even the most contradictory word, preserves contact—it is silence which isolates." (Thomas Mann: *The Magic Mountain:* Chapter 6)
6. Socrates was put to death for his iconoclastic views. We hope the reader will not be as harsh.
7. In order to calculate the information transmitted per phoneme it is necessary to consider all of the possible alternatives, including all of the phonemes of the language as well as silence.
8. Interaural phase is also an important consideration in binaural speech perception.
9. "The game's going on rather better now," she [Alice] said, by way of keeping up the conversation a little. " 'Tis so," said the Duchess: "and the moral of that is—'Oh, 'tis love, 'tis love, that makes the world go round!' " "Somebody said," Alice whispered, "that it's done by everybody minding their own business!" "Ah well! It means much the same thing," said the Duchess, digging her sharp little chin into Alice's shoulder as she added, "and the moral of *that* is—'Take care of the sense, and the sounds will take care of themselves.' " (Lewis Carroll: *Alice's Adventures in Wonderland,* 1865, Chapter 9)

REFERENCES

Boothroyd, A. Statistical theory of the speech discrimination score. *Journal of the Acoustical Society of America,* 1968, *43,* 362–367.

Bronstein, A. J. *The pronunciation of American English.* Englewood Cliffs, N.J.: Prentice-Hall, 1960.

Denes, P. B. On the statistics of spoken English. *Journal of the Acoustical Society of America,* 1963, *35,* 892–904.

Dubno, J. R. & Levitt, H. Predicting consonant confusions from acoustic analysis. *Journal of the Acoustical Society of America,* 1981, *69,* 249–261.

Erber, N. P. Auditory, visual, and auditory-visual recognition of consonants by children with normal and impaired hearing. *Journal of Speech and Hearing Research,* 1972, *15,* 413–422.

Fant, G. *Speech sounds and features.* Cambridge, Mass.: M.I.T. Press, 1973.

Gabor, D. Theory of communication. *Journal of the Institution of Electrical Engineers,* 1946, *93,* 429–457.

Green, D. M., & Birdsall, T. G. The effect of vocabulary size on articulation score. *Technical Report #81,* Engineering Research Institute, University of Michigan, 1958.

Jacobson, R., Fant, G., & Halle, M. *Preliminaries to speech analysis: The distinctive features and their correlates.* Cambridge, Mass.: M.I.T. Press, 1951.

Kalikow, D. N., Stevens, K. N., & Elliot, L. L. Development of a test of speech intelligibility in noise using sentence materials with controlled word predictability. *Journal of the Acoustical Society of America,* 1977, *61,* 1337–1351.

Kewley-Port, D. & Pisoni, D. B. Discrimination of rise time in nonspeech signals: Is it categorical or noncategorical? *Journal of the Acoustical Society of America*, 1982, *71*, S36 (Abstract).

Klatt, D. H. Linguistic uses of segmental duration in English: Acoustic and perceptual evidence. *Journal of the Acoustical Society of America*, 1976, *59*, 1208–1221.

Levitt, H. Transformed up-down methods in psychoacoustics. *Journal of the Acoustical Society of America*, 1971, *49*, 467–477.

Levitt, H. Acoustic analysis of deaf speech using digital processing techniques. *IEEE Transactions on Audio and Electroacoustics*, 1972, *AU-20*, 35–41.

Miller, G. A., Heise, G. A., & Lichten, D. Intelligibility of speech as a function of the context of the test material. *Journal of Experimental Psychology*, 1951, *41*, 329–335.

Miller, G. A., & Nicely, P. E. An analysis of perceptual confusions among some English consonants. *Journal of the Acoustical Society of America*, 1955, *27*, 338–352.

Osberger, M. J. A comparison between procedures used to locate segment boundaries. *Journal of the Acoustical Society of America*, 1979, *65*, S66 (Abstract).

Peterson, G. E., & Lehiste, I. Duration of syllable nuclei in English. *Journal of the Acoustical Society of America*, 1960, *32*, 693–703.

Pollack, I. Message probability and message reception. *Journal of the Acoustical Society of America*, 1964, *36*, 937–945.

Shannon, E. C., & Weaver, W. *The mathematical theory of communication*. Urbana, Ill.: University of Illinois Press, 1949.

Walden, B. E., Prosek, R. A., & Worthington, W. W. Predicting audiovisual consonant recognition performance of hearing-impaired adults. *Journal of Speech and Hearing Research*, 1974, *17*, 270–278.

16

Recipe for Relevance
Latin and Its Literature

Samuel Lieberman

Late of Department of Classical and Oriental Languages, Queens College of the City University of New York

One reason for calling Classical literature Classical is that it is supposed to be always relevant, meaningful, regardless of the historical period in which it is read and studied. And the reason that this literature, Roman and Greek, has not only survived over two and a half millennia but has served as source, model, and inspiration for later literatures and ideas is that this supposition has turned out to be true. Yet somehow the educated public of the United States today is more willing to see the relevance of Greek literature than that of Roman, especially in the matter of war and peace in the currently highly fashionable theme of love, which is so "relevant" that in the words of a song popular about a generation ago, it "is busting out all over." This must be recognized as a serious problem for Classics teachers since most of them teach Latin, if they are still in the field in this era of dwindling enrollments in foreign languages when many young people are beginning to turn away even from modern languages. The problem, I submit, is in part of our own making and in part the result of related educational and cultural history.

I have recently been lecturing to various groups, the general public as well as Classics teachers, on the subject of "Roman Poets who Sang: Make Love Not War," and my audiences have been invariably surprised to find that writers who were *Romans* could be so sensitive, tender, and insightful about love (even the classics teachers who had been exposed to Catullus) and so expressive about their opposition to that activity for which the Romans allegedly had such a love, war. Also, it is the common experience in colleges and schools that courses in Greek literature in translation are extremely popular, whereas courses in Latin literature in translation, where and when they are given separately, are far less attractive. For this Classicist, who has dealt with both literatures in the original and in translation, this requires an explanation, and for teachers of Latin, who may need to supplement their dwindling language classes with courses in literature in translation and in Classical civilization, this is a serious challenge and, to be brutally frank, a threat. An explanation for the lower opinion of Latin as compared to Greek literature is, as I shall show, available and, if it is correct, may help supply the means to meet the challenge and remedy the situation. Quite clearly the Romans have had a "bad press" and I think the reason is two-fold, each revealing paradoxical elements.

To begin with, for the past two generations in the United States, as the study of classical languages, particularly at the secondary school, declined together with the virtual disappearance of ancient Greek at that level, many more people studied Latin than Greek and more have been exposed to bits and pieces of Latin literature in the original than Greek in the original. And the bits and pieces they had to read were studied more for grammatical and linguistic analysis than for their meaningfulness to the reader, their esthetic qualities, and their bearing on the student's life and current realities. On the other hand, the average person exposed to Greek literature has read it only in translation, reading whole works through, and classroom discussion has centered on the ideas, the characters, their inner motivations as well as the impact of their societies on them, and on the meaning of the work in terms of the realities and needs of the current world and its ideas. So Greek literature seemed to live and stimulate readers while Latin literature seemed just to lie there ponderously as raw material for grammar, syntax, and translation into tortured English.

The other reason is an out-and-out paradox from which arose a faulty image. The Roman image suffers from the success of Roman civilization! And since Roman success seemed to lie in their efficiency in war, government, law, and administration, the Roman image has come down to us of a people as soldiers, administrators, and lawgivers, usually quite dour and stern (except for occasional rumored bursts of sadomasochistic "orgies"). In the popular view, even of the educated, they are a rather stodgy, unimaginative people not given to flights of ideas or fancy like the Greeks. Who expects sensitivity, wit, delicacy, imagination of a Roman? (Few people even see them as the Italians they were, which would conjure up a different stereotype!) When a Roman does display such qualities, there has until recently always been some scholar to attribute it to Celtic blood in his veins even though when the Romans were active, powerful, and writing there is no evidence of so-called Celtic fire in the then barbarian Gauls or Galatians whom they fought and absorbed.

Vergil himself, in the very act of producing one of the world's greatest pieces of imaginative literature, was so modest as to contribute to this unfortunate Roman stereotype. In *Aeneid* VI, 847–853, he puts into Anchises's mouth the famous lines prophesying Rome's political mission to the world conceding the arts and sciences to the Greeks (the "others" in the prophecy):

> Others will beat out bronze into living shapes more artistically,
> and carve living features from marble,
> plead cases better, and describe and measure the motions of the heavens,
> and explain the rising and setting of the stars.
> You, Roman, remember your mission:
> to rule the peoples by your sway
> —these will be your arts and sciences—
> to impose the habits of peace,
> to spare those subject to you,
> to beat down the stubbornly bellicose.

But this is not the only theme of this great Roman epic poet of humanism and humanness, and it is our responsibility, our mission, as classics teachers, to bring this out despite Vergil's modesty, whether teaching it in translation or in the original.

In such matters, therefore, as war and peace, or love and hate (despite Catullus' brief but insightful and intense *odi et amo,* Poem 85, anticipating Freud as well as many other poems), we are willing to accept that Greek writers could produce works that speak against war but tend to accept as certain that this was not what Romans ever wrote against. We think immediately of Greek plays like *Lysistrata, Peace, Trojan Women,* and can even extract some antiwar sentiment in the warlike *Iliad.* But the Romans? Works by them on this theme do not come so readily to the minds of most people. Yet they too had poets who spoke against war and even poets whose whole subject was the now popular theme of "make love not war." These were the leading poets of the later Augustan era: Propertius, Tibullus, Ovid, for the first two of whom it was their chief theme.

The two leading poets of the first half of Augustus' principate, Vergil and Horace, were no lovers of war, but both supported the restored security and peace of the new Augustan order and in praising it inevitably extolled its military might (as in the passage quoted above). Both frequently expressed abhorrence of war, especially civil war, through which they lived and suffered. Thus the *Aeneid*, the poem of Roman national ideals and *lacrimae rerum*, is filled with the tragedy of inescapable war in which fine young men on both sides have to kill one another so that peace may be established. And Vergil's *Fourth Eclogue*, written in the midst of civil war, is a poem of passionate longing for a time when peace will come, symbolized by the birth of a beloved and loving child. Horace in two Epodes, VII and XVI, bitterly assails civil war. But both these poets are spokesmen for the new regime and extol its virtues, which they see as ensuring peace and security after two generations of bloody internecine strife.

It is with the younger generation of the Augustan period that antiwar sentiment becomes a dominant theme together with the seemingly un-Roman refusal to serve in any army and extolling love, especially unmarried love, between a man and woman as the only virtue. They *literally* sang, "Make love, not war." Propertius in *Elegies* II, 7.14 writes: *nulles de nostro sanguine miles erit* ("No one from my blood will be a soldier"). Tibullus in *Elegies* I, 1.76 writes: "It is in bed that I am a good captain and soldier" *(Hic ego dux milesque bonus)*. And in his *Elegies* I, 10.47–48 we read: "Peace tends the vine and stores its juice/so sons may live to drink their fathers' wine" *(Pax aluit vites et sucos condidit uvae/funderet ut nato testa paterna merum)*. Ovid, *Amores* I, 9, extols the lover as being as good as if not better than the soldier in the risks, dangers, and rewards he gets from his activities and in the tactics and strategies to which he resorts to attain victory. There is much more by these poets, complete and masterly poems on this extremely timely theme for the Vietnam generation and its successors. It is no wonder that the audiences to whom I spoke were so gratifyingly surprised when this was presented to them and in much greater detail. The Romans turned out to be human after all. But to be aware of it and of the material illustrating it, to get the point across to our students and have them read and understand the material is our job.

This currently popular theme is of course not the only humanistic or meaningful one that Roman, like Greek, literature deals with. Let me suggest a few works outside the standard secondary-school curriculum for second- or third-year Latin (with the hope that a successful second-year class will help retain or restore a third or even fourth year). Our Latin students do not have to spend all their energies slogging as footsoldiers through the wilds of Gaul or haranguing the Roman senate or courtroom. Let me add here that suggestions for broadening the range of authors to be read in such Latin classes are nothing new. They have been made for several years and have been incorporated in the published curricula of a number of state departments of education[1] but have usually not been carried out in practice, whatever the reasons. Enough of preliminaries; let us get to the substance.

Cicero's philosophical works, which discuss primarily social and ethical problems and are for the most part written in a Latin no more difficult than that of his speeches, should be most interesting to today's students and would stimulate much fruitful class discussion. A very good, relatively new, inexpensive school edition of selections with useful notes, commentary, and vocabulary has been available for some time.[2] Terence's plays should be very successful on a secondary-school level. A second- or third-year student should be able to handle this language as competently as Caesar's, and his themes are "a natural" for young people (or their elders): the generation gap, the clash between fathers and sons, love between young people disapproved by their elders (who in some cases turn out to be hypocrites). In offering Terence to students after they have mastered the elements of Latin we are supported by an old precedent, for that was the stage at which this author was taught several hundred years ago before the now receding reticences of Victorianism relegated him to more mature readers at the college level. In addition, in Terence students will have the refreshing experience of being exposed to the liveliness of good colloquial Latin, rarely met today below college level.

One could even try in third-year Latin one or two of the essays of Seneca. The choppy epigrammatic style, reminiscent of Carlyle's and of some American writers of the last two decades, may be a bit startling at first, but with practice the students, like their teachers, will find that Seneca's writing is only a kind of intensified conversation exploring themes of great appeal for the aware student. Chief among them is the Stoic stress on the prevalence of reason in all human beings and the consequent right of all, regardless of class or origin, to be treated as rational and essentially equal with certain innate human rights. Students can be made to see in such Stoic doctrines the beginning of the powerful concept of the Rights of Man which has played such an important role in the civilized world at least since the late seventeenth century. Or they could try Seneca's hilarious *Pumpkinification of the Deified Emperor Claudius (Apocolocynthosis Divi Claudii)* just for fun.

There is much else in Latin Literature of great writing, stirring themes, or sheer enjoyment, and of continuing modern significance. For those who teach literature in translation there are in addition (though read in the original usually only on the college level) biting satire (such as the Greeks never left us), fine lyric and elegiac poetry, trenchant history, novels, and, of course, Vergil's great epic and other poetry, and Ovid's many works, and much more. There is no reason to undervalue Roman literature. In its way it is as great as the Greek and in some genres the Greeks left nothing comparable. Regardless of nonextant but possible Greek predecessors, Roman satire, love-elegy, the two novels: *The Golden Ass* and *Satyricon* are uniquely Latin. It is the Classics teacher's responsibility, whether teaching Latin or Greek, to remember that he or she is dealing with a humanistic literature,[3] that the ideas and ideals expressed in this literature, though started by the Greeks, were continued and further developed by the best minds among the Romans, and that it was through their Latin version that they first had their tremendous impact on the Western world. Greek philosophy influenced Western thinking for hundreds of years through the writing of Cicero, and in the sixteenth- and seventeenth-century revival of drama it was the plays of Plautus, Terence, and Seneca that served as models rather than those of the Greeks, which had served as the Roman dramatists' model and source.

It is up to us teachers to make our students and the public aware of all this, to think and teach creatively, and to innovate where necessary, realizing that much of our innovation may even be a kind of revival. Hopeful signs at the school level are the successful new classics programs in the "inner-city" schools of Washington, D.C. and Philadelphia, which have been reported extensively.[4] To call once more on that old cliché, there is no need to curse—and wail—in the dark. Light a few candles on your own and they may turn out to be *Roman* candles, fireworks!

NOTES

1. For example, the latest New York State Syllabus for Latin, *Latin for Secondary Schools,* recommends the reading of selections from the following authors at level II: Bede, Catullus, Cicero *(De Sen.),* Eutropius, Florus, Gellius, Livy, Martial, Nepos, Sallust, Velleius Paterculus, as well as adaptations from Plautus and Terence and selections from Medieval and Renaissance prose and poetry. Recommended for Level III, pp. 38–39, are (in addition to Caesar and Cicero): Florus, Gellius, Livy, Nepos, Pliny the Younger, Quintilian, Sallust. A similarly wide range of authors is recommended (pp. 48–49) for Level IV besides the *Aeneid* including, for example, Horace, Propertius, Vergil's *Eclogues,* and Ovid's *Metamorphoses* (the latter I know teachers of fourth-year Latin have always read if they were fortunate enough to get a class at this level). This syllabus also recommends organizing the reading and the units around topics and themes characteristic of Roman civilization and thought. The Latin syllabi of other states will, I am sure, prove to make similar recommendations.
2. This text, Wilson (1964), is available in the United States. For other texts available consult either *Classical World* or *American Classical Review*

(ACL), each of which publishes in two separate issues each year lists of Latin and Greek texts available for school or college use and lists of paperbacks in English on classical subjects. State syllabi for Latin, such as that of New York referred to in note 1, also supply such information, although on a much smaller scale.

3. For a more extended treatment of the need to emphasize the humanism in Latin, Greek and other Classics courses, see Lieberman (1971).
4. For Philadelphia, see Masciantonio (1971). For Washington, D.C., see LeBovit (n.d.). In his annual *ACL Newsletter*, John Latimer has also briefly reported on both of these programs. Further information can be obtained by writing to the Education Departments of both these cities.

REFERENCES

Latin for secondary schools. Albany, New York: The University of the State of New York, The State Education Department, Bureau of Secondary Curriculum Development, 1971, pp. 14–15.

LeBovit, J. *The teaching of Latin in elementary schools.* McLean, Va.: Latin for the Modern School, Associates n.d.

Lieberman, S. The humanities as human studies. *Classical World,* 1971, *64,* 262–263.

Masciantonio, R. The implications of innovative classical programs in the public schools of Philadelphia. *Classical World,* 1971, *64,* 263–264.

Wilson, S. J. *The thought of Cicero.* London: G. Bell, 1964.

17

The Speech of New York City
The Historical Background

Raven I. McDavid, Jr.

Professor Emeritus, Departments of English and Linguistics, University of Chicago and Editor-in-Chief, Linguistic Atlas of the Middle and South Atlantic States

In this tribute to Arthur Bronstein, one of the most distinguished students of present-day New York City speech, it is appropriate that one examine the earliest historical roots of that speech. Lest someone demur that my recent Carolina upbringing disqualifies me from competence for the task, I must observe that both of my maternal grandparents were natives of New York State: my grandfather was born in Oswego, my grandmother in Setauket, Long Island, where her parents were refugeeing from the summer heat of lower Manhattan. Her connections with the city go back at least to 1637, with the arrival of Oloffe Stevensen, whose descendants assumed as a surname his soubriquet Van Cortland, probably derived from their experience at trading in the Baltic. My last ancestor

Originally presented at the 1977 meeting of the National Council of Teachers of English, in New York City, this historical study, for the *Handbook of the Linguistic Geography of the Middle and South Atlantic States,* was made possible by a Senior Fellowship from the National Endowment for the Humanities and by the hospitality of the Newberry Library of Chicago and the Library of the New York State Historical Association at Cooperstown. Linguistic evidence is principally derived from the field records for the Linguistic Atlas project, inaugurated by the American Council of Learned Societies and directed by Hans Kurath.

to arrive in this hemisphere, Jan Cornelis Van Den Heuvel, intermarried with the Apthorps of somewhat notorious reputation; he had a town house at 229 Broadway and a country retreat in Bloomingdale, now the west seventies.[1] Others, with varied professions and interests, lived in Long Island, Brooklyn, Manhattan, and the Albany area—an interesting crew.[2]

The place of New York City speech in American dialects has long been debated. It is not to be confused with the speech of eastern New England: its prestigious "broad a" [ɑ] is of a different phonetic quality and distribution from that of *Harvard Yard* [a]; it never had the "New England short *o* [ϴ] in *coat, road, home;* it has consistently simplified the initial cluster in *whip,* and followed close behind the Received Pronunciation of London in neutralizing the contrast between such pairs as *horse/hoarse, morning/mourning.* Still unexplained is the fact that *birds chirp* (or did) with [3ɪ] in New York City, even in the most refined society—in a fashion unknown in any dialect of the British Isles, though not unknown in plantation areas of the Gulf States; and that many New Yorkers, including numbers of the élite, never distinguished a *curl* of hair from a *coil* of rope.[3] Some of these features are now matters of linguistic history; but it is as interesting to speculate what lay behind that past model as it is to see what is happening to the model under the impact of new migrations and recent social changes.

Before the days of exurban sprawl, New York City speech was tolerably well confined, unlike other distinctive urban varieties of Atlantic Seaboard speech. Boston forms are found as far west as Rochester, among the educated, and to some extent are affected by the élite of such inland areas as Chicago (Uskup, 1974). Virginia plantation forms are solidly established up to the Blue Ridge and emulated further west; Charleston speech dominates the South Carolina Low Country, with outposts at Augusta, Georgia, and in the lower Piedmont. But most New York speech forms stop with Westchester County, the Jersey Meadows, and Oyster Bay. We need to look at linguistic details in the light of settlement and other developments.

A few historical tidbits show that things have changed less than some spokesmen would assert. Not merely in election years, city politicans have been asking special favors from the nation since 1776. When the British fleet dropped anchor in the Lower Bay, George Washington recognized that the city was indefensible in the face of overwhelming seapower, and wished to withdraw his army intact. His military judgment overruled, he suffered crushing defeats, with disastrous losses of men and matériel, and almost the loss of the Revolution itself. Unhappy relationships with the rest of the state are an old story: Tammany voted against the Erie Canal, which more than anything else contributed to the commercial preeminence of the city. For the first 40 years after national independence, the mayor of New York was appointed by the governor. In the 1870s, despairing of local corruption, the legislature created a metropolitan police force for New York and Brooklyn; when the local police would not bow out, there came a genuine police riot, put down by the militia. The charter for Greater New York was never approved by the voters, but imposed by the legislature, after successive vetoes by the mayors of Brooklyn and New York; the annexation of Flushing and other communities in Queens was in effect an act of legislative rape. The Port Authority is but a feeble effort to rectify the mistake of the Stuarts in dividing the harbor between two provincial jurisdictions, contrary to the better example of the Dutch. No wonder the city sometimes toys with secession from the state: in 1788 it was seriously proposed that a seceding metropolitan area might ratify the Federal Constitution if Upstate continued to demur; and in 1860, after the city had given a 30,000 vote majority against Lincoln, Mayor Fernando Wood suggested it become an independent free-trade city, much like Hong Kong today, to insure its profitable commerce with the South.

Nor has New York lacked its interesting administrators, such as the succession of royal governors after 1688 (we can pass over the Dutch Van Twiller and Kieft, and the mayors since 1865). Henry Sloughter, the first of these, under the influence of strong drink, signed the warrant for hanging and quartering his predecessor Jacob Leisler, the first New York reform administrator and the first administrator to come to power with popular support. The second, Fletcher, had so profitable an arrangement with the Red Sea pirates that when his successor, Lord Bellomont, tried to stop the traffic, he was assailed for hurting business. The fourth, Lord Cornbury, who frequently appeared in drag, was perhaps the first closet queen to come out in America.

There have been enough spontaneous demonstrations, as they are now called, to fill several books. The two alleged "Negro plots" of 1712 and 1741 saw more blacks burned alive—with judicial sanction and public approbation—than in the entire history of Mississippi. The roster of subsequent demonstrations includes the Doctors' Riot (a medical school wrecked for alleged grave-robbing), the Flour Riot (protesting high prices and destroying much of the supply during the protest), the Stonecutters' Riot (protesting unfair competition by the guests of the state at Sing Sing), the Theater Riot (provoked by two competing productions of *Macbeth* by the Briton Macready and the American Forrest), the great Draft Riots of 1863 (finally quelled by Federal troops, with the death toll never officially tallied but estimated as high as 3,000), and the Orange Riot, when a group of Protestant Irish asserted their constitutional right to parade in memory of The Boyne, and paraded with a permit and the protection of Catholic Irish police who shot to kill other Catholic Irish staging a counter-demonstration.

Against this we must weigh a history of foresighted social legislation: the first city government north of Mexico, the first municipal charter, religious toleration, early public schools,[4] respect for learning (Cadwallader Colden, the last royal lieutenant governor, was of intellectual stature comparable to that of Benjamin Franklin), a library system, private and public colleges, art galleries and museums, generous and imaginative philanthropy. Truly a complex city.

The complexity goes back to its origins. The harbor was discovered by Verrazzano, an Italian in French service; the harbor and the river were explored by Hudson, an Englishman under Dutch orders; the first settlements—Albany in 1624 and Manhattan two years later—were under the Dutch flag; and Peter Minuit, who arranged the purchase of Manhattan, later founded the rival colony of New Sweden on the Delaware.

As befits a mercantile plantation—and New York has never fed itself—even the earliest settlements were mixed. Nearly half the Netherlanders were French-speaking Walloons, refugees from the southern provinces, which had remained Roman Catholic and loyal to Spain; their origins are reflected in Wallabout Bay.[5] Not only did English-speaking settlements dominate Long Island from Oyster Bay east, but there were several such settlements within the present corporate limits of New York City, reflecting distaste for the blessings of New England theocracy—Ann Hutchinson's colony in the present Bronx, Gravesend in Kings, Newtown and Flushing in Queens. By the 1640s the Jesuit martyr, Father Jogues, reported that 18 languages and dialects were spoken in New Amsterdam. Leisler was a German; Sephardim migrated from South America when the Portuguese reconquered the Dutch lodgment in Brazil; blacks were introduced as house servants and craftsmen by Dutch slavers. The hint of Slavic things to come was found in Prague Germans, one of whom left New York in the midseventeenth century to set up the Bohemia Manor in Delaware. And well before the Revolution there were Irish from north and south.

Unlike most colonial towns, New York never had a serious agricultural hinterland, though it was the first American community to establish standards for domestic flour. It has existed primarily for trade, with a seldom matched harbor and with the best route inland to the North American heartland. The Dutch came for the fur trade and made satisfactory arrangements with the Iroquois, the Cosa Nostra of the eastern Indians, arrangements the English continued to their profit. When the fur trade played out, the city became the center for piracy; then for half a century it was the chief military base for the British against French Canada, with successful privateering on the side. Trade resumed after the Revolution, chiefly in the North Atlantic, but with greater profits for longer ventures; with its network of coastal companies, New York was also the center for shipping cotton to Europe and for imports to the cotton states. It introduced the first liners—freighters sailing on a regular schedule—and the liners that carried cargo east brought immigrants back. Even though the British were more successful in introducing steam navigation, they continued the pattern of traffic from Europe to America through New York. By 1820 the city had become the preeminent American port. The Erie Canal and the railroads extended its trading area across the continent.

An active port enriches its merchant class; and New York fits the tradition. Its merchants dominated the city from the Eighteenth Century. Though there have been spectacular examples of new money, more important is the continuity of the old merchant class in positions of influence. Down to the development of the railroads, the New York business community largely grew from within, with small accretions from outside. There were some New Englanders—but few from Boston, except as Bostonians like the Apthorps set up branches of their firms, as was done to a lesser degree by Charlestonians and New Orleanians. New York City was spared, for the most part, the great Yankee invasion that swept across Upstate and on to the Pacific. Its mercantile élite, the models for social behavior including speech behavior, were indigenous. This was likewise true of the expansion of the mercantile class into banking, insurance, and securities dealing. Despite the movements of population and industry, seven of the ten largest American banks are still located in New York City. Some have interesting histories: the Manhattan Company was originally chartered to supply drinking water, and it kept the pumps operating—for fear of losing its charter—till the building of the Croton Aqueduct. The New York Stock Exchange and several insurance companies also were in operation before 1800.

Merchants and bankers are basically conservative, disinclined to disturb the *status quo,* inclined to hedge their bets, but not disinclined to profit from disturbances which others have created. Down to 1860, New York continued to finance the slave trade, largely for the Cuban market, but occasionally for illicit importations to the cotton states. Like the Scots families in the troubled Eighteenth Century, New York businesses protected themselves by having prominent partners involved in both major political parties, so that their access to state contracts would be unimpeded by changes in administration. The one exception was 1914, when the heavy commitment of business houses to one side made a mockery of Wilson's carefully planned neutrality. Perhaps this experience lies behind the slowness of some financial houses to espouse passionately touted social causes: a sense of history can make one hesitant.

A mercantile and financial center is not without its disadvantages. Its power to make money rapidly is matched by a similar power to lose it; gains on risk capital can be wiped out overnight, and in times of delicate adjustment the troubles of the New York business community affect not only the rest of the city but the nation—even the world. Almost every American depression or panic has started from troubles in the New York business and financial community. Furthermore, when a city thrives on trade and finance, there are many who do not share its prosperity, even in the flushest of times; as far back as the 1820s there were complaints about the high cost of poor relief. And the generosity of New York in dealing with its own poor has always attracted them from elsewhere. The grimness of their fate contrasts with the conspicuous waste of the "beautiful people"—an old story in a city where poverty and affluence have long marched cheek by jowl. Only occasionally, as with Hazen's three-act Versailles dinner at Sherry's a century ago, when the publicity forced his removal as an insurance executive, has public indignation checked this extravagance. For the most part, what the rich and super-rich do is of a piece with other forms of entertainment, such as that furnished by the Yankees, the Mets, the Islanders, the Rockettes, the Rangers, the Knicks, and the oft-lamented Dodgers. Very few members of the old New York élite indulge in such extravagances; they are generally content to go their way, assimilating such newcomers as the Astors, the Vanderbilts, the Belmonts, and the Rockefellers, and transmitting their models of speech and other decorum. This makes for continuing stratification in speech, with possibly striking new patterns of distribution of the strata; but such new distribution is a familiar story to the student of human mores.

Now to return to the local prestige model—however obsolescent it may be—its sharp differences from the prestigious speech of nearby areas, and the ill feeling between the City and Upstate: we may find the roots in the Civil War of 1775–1783, as it was frequently called on both sides of the Atlantic. As a trading center, New York was conservative; the merchants objected to the Stamp Act and other new taxes but were reluctant to make an overt break with England. Loyalist strength in the colony was about half, somewhat greater in the New York City area; at most, 30% were firmly committed to the Continental cause; in Westchester it was difficult to raise Continental militia.

Throughout the troubles, there were dramatic shifts in the population of the city, reflecting the tides of war. As early as 1774 there were the beginnings of Loyalist flight; in 1776, with Howe's fleet poised in the Lower Bay, adherents of both sides found it convenient to leave.[6] From a 1775 population of some 25,000, the city had shrunk to less than 3,000 by the time the British established control; one British observer said that 95% of the population had fled.

But the city was largely intact, despite fires in 1776 and 1778—not ordered by Washington, though suspicion provided a few hangings for public entertainment. It was still an important and secure port; it became the headquarters of the North American Expeditionary Force, with tremendous military and naval traffic. Loyalists drifted into the city, joining the British garrison, which itself reached 10,000 men, with some 25,000 military dependents. There were hardships: the city's forests were cut down to provide firewood; even Delancey's Westchester Refugees and the less respectable Cowboys (ostensibly irregular Loyalist militia, but in reality an early example of local supply-side free enterprise) could not procure enough local food for the city, and supply convoys at sea were subject to privateering from elsewhere (New York continued active as privateer center for the British). But life was comfortable; living for the privileged was gracious, with witty British officers in attendance at social functions— notably the Adjutant-General John André, whose ill-managed communication with the traitor Arnold was to cost him his life. And for seven years there were no taxes. At no period has the influence on New York City of British manners and mores—including British speech—been more intense. With the continuing influx of Loyalists, the population contained a growing majority of those sympathetic not only to the King's cause but to the King's English. The withdrawal of all but one garrison from Westchester County in 1779 served only to widen the ironically titled Neutral Zone, to encourage the pillaging of the local population by both sides, and to intensify the difference between city and state.

This came to an end with the great evacuation of 1783, when from New York port departed not only soldiers and dependents and the civil government they had maintained, but some 40,000 Loyalists, possibly half from New York State. The population shrank again, to some twelve thousand, but filled up again to thirty thousand by 1790.

But it was a different city from the colonial capital of 1775. Probably more local Loyalists remained than left; and of the latter, some ultimately came back. With its traditional pragmatism, New York did not engage in a serious purge. Government and business must go on; a port is dependent on trade, and if it loses all those who understand trade, it will collapse. There also seems to have operated another kind of pragmatism: a recognition that many Loyalists were in their way just as sensitive to human rights as the Continentals, and had as often made their choice as a hairline matter of conscience. The spirit of accommodation did not prevail in Westchester, nor in the Schoharie Valley where the Iroquois had ravaged the frontier settlements; but New York City, as we know, was different.[7]

And the occupation had linguistic effects. For seven years the present-day New York Metropolitan area had been under British control; it had been cut off from its hinterland by a vaguely defined neutral area across which both sides raided but over which neither established effective control. Bloody skirmishes took place around the perimeter, but the city remained firmly in British hands, subject to British mores and infiltrated by British speechways. Continental resentment Upstate, at the comforts enjoyed by the Loyalists in the occupied city, has been one of the major strands in Upstate suspicion of the city and rejection of what it stands for. But likewise the impress of the occupation lingers, as attenuated as it may be by recent population movements. Perhaps I am too much of a historian and traditionalist, and liable to exaggeration; yet the case is put by maps in Kurath's (1949) *Word Geography,* by Frank's (1949) dissertation, by Van Riper's (1958) study of postvocalic /-r/, and by other studies derived from the collections of the regional linguistic atlases. Even if it is pure coincidence—and I suspect there is more—the fact remains that the limits of the area of New York City speech, as delimited by Kurath, remarkably parallel the limits of effective British control from 1776 to 1783.

NOTES

1. The career of Charles Ward Apthorp, Van Den Heuvel's father-in-law, is an interesting illustration of the vicissitude of the New York élite in the Eighteenth Century. Of a successful Boston mercantile family (his father was the agent for the Massachusetts colonial government who arranged for the transportation of the Acadians out of Nova Scotia), he had prospered in business, land speculation and politics; during the French and Indian War he was paymaster of the British forces in North America. A member of the Council of the City of New York he decamped from Manhattan on the eve of the British invasion; during the occupation he served on the Governor's Council. Although he lost some of his lands—on the Kennebec, in Connecticut, and in New Jersey—and his Boston mansion on Beacon Hill, he saved his social position and most of his fortune, partly by advantageous marriages of his daughters, one to Van Den Heuvel and another to Hugh Williamson, a signer of the Constitution from North Carolina.
2. Many of the family papers are currently in the possession of my cousin William H. Swan, of Quogue. Those dealing with the Apthorps have been catalogued by my sister, Mrs. L. L. Barrett, and are on loan to the North Carolina archives in Raleigh.
3. Knowing this as a social shibboleth, I was surprised when I first encountered this homonymy (c. 1930) in the speech of a corporation executive, of New York City Dutch Colonial stock.
4. Providing public schools and patronizing them were different matters; even before the recent immigration from the South and the Caribbean, it was a rare child in Brooklyn Heights who attended public schools.

5. About 1800 it was observed that of the inhabitants of Staten Island a third spoke French, a third Dutch, a third English.
6. Several prominent merchants, including Charles Ward Apthorp, seem to have sought hospitality (and received it) on shipboard before the first British landing in the Lower Bay.
7. The proportion of blacks in the population of 1786 is strikingly lower than in 1771, since many had either departed with their Loyalist owners or been carried away as booty by the departing British, to the Maritimes or the Bahamas. In 1771 there were 10,559 blacks of a total population of 60,696 in New York, Kings, Queens, Richmond, and Westchester counties; in 1786, though the total population had risen slightly to 64,390, the number of blacks had sharply declined, to 7,546.

Other coastal areas showed a similar loss of black population. The 1775 South Carolina population of 200,000 was 62.5% black; the 1790 population of 240,673, slightly more than 40.0%, with fewer blacks than 15 years earlier. Restoration of the plantation labor force to its pre-Revolutionary size was one of the motives behind the Constitutional provision forbidding interference with the slave trade till 1808. By 1810 South Carolina once more had a striking black majority, a situation that lasted till 1920; the new importations contributed heavily to the establishment in the plantation area of Gullah, a unique creolized dialect.

REFERENCES

Avis, W. S. The New England 'short o': A recessive phoneme. *Language*, 1961, *37*, 544–558.
Frank, Y. H. *The speech of New York City*. Unpublished Ph.D. Dissertation, University of Michigan, 1949.
Hubbell, A. H. *The pronunciation of English in New York City: Consonants and vowels*. New York: Kings Crown Press, 1950.
Kurath, H. *A word geography of the eastern United States*. Ann Arbor : University of Michigan Press, 1949.
Kurath, H., & McDavid, R. I., Jr. *The pronunciation of English in the Atlantic States*. Ann Arbor : University of Michigan Press, 1961. (Reprinted Alabama, University of Alabama Press, 1983.)
Labov, W. *The social stratification of English in New York City*. Washington, D.C.: Center for Applied Linguistics, 1966.
McDavid, R. I., Jr. The folk vocabulary of New York State. *NY Folklore*, 1951, *7*, 173–191.
Uskup, F. L. *Social markers of urban speech: A study of elites in Chicago*. Unpublished Ph.D. Dissertation, Illinois Institute of Technology, 1974.
Van Riper, W. R. *The loss of postvocalic /r/ in the eastern United States*. Unpublished Ph.D. Dissertation, University of Michigan, 1958.

18

Language and Psychoanalysis

Donald Meltzer

London and Oxford, The United Kingdom

THE LINGUISTIC PROBLEM IN DREAM STRUCTURE AND AUTISTIC MUTISM

In this paper I should like to set out clearly the problems present in three areas—linguistics, dream structure, and autistic mutism—and, by tracing the implications of their interaction, throw some light on all three. The problem of linguistics, or psycholinguistics as it is now called, which concern us are of two sorts, at two ends of a methodological spectrum. At one end is the mind–body problem dogging the footsteps of behaviorist psychologists like Skinner and mathematical linguists like Chomsky and Katz. At the other end is the problem of cosmic mysticism, the expression of which, in the spirit of Jung or Ouspensky, is the Whorf–Sapir theory of the relativity of language. I will discuss these in order to define the position of psychoanalysis, methodologically, and also to stress those mysterious aspects of grammar and semantics which link with our other two areas, dreams and autism.

After this linguistic preamble, I will then turn to the problem of dream structure, its relation to waking thought and to inner and outer speech. For the purpose I will use some clinical material from a young poet and from a gifted psychotic girl.

Following this excursion into the syntax of the dream, I will turn to our recent findings in studies with autistic children to state a theory of the nature of mutism in these cases, comparing them with catatonic

169

mutism. With this theory in hand, I will return to dreams and language development in general, ending with a discussion of Wittgenstein's theory of language, or rather, his penetrating questions about language.

The Mind–Body Problem

All linguistic studies start with the assumption that language as spoken is composed of basic units which are arranged inside the head and emitted by the oral apparatus. This is common sense and self-evident, but I will argue that it is not correct, from the psychoanalytical point of view. The great cleavage in the linguistic field is over the issue of whether the primary unit of language is the phoneme—the unit of sound—or the morpheme—the unit of meaning. Those who oppose the phoneme as an artificial unit created by phonologists, on whatever ground they base their argument—psychological, methodological, or theoretical—seem nonetheless to assume that a *unit* must be found.

The reason for this is fairly apparent, for all linguists appear to assume that language has the primary function of communicating information. For example, to quote from Jakobson and Halle (1955):

> The addressee of a coded message is supposed to be in possession of the code and through it he interprets the message. Unlike the decoder, the cryptanalyst comes into possession of a message with no prior knowledge of the underlying code and must break this code through dextrous manipulation of the message. A native speaker responds to any text in his language as a regular decoder, whereas a stranger, unfamiliar with the language, faces the same text as a cryptanalyst. A linguist, approaching a totally unknown language, starts as a cryptanalyst until, through a gradual breaking of its code, he finally succeeds in approaching any message in this language like a native decoder.

It is apparent that this is not a model, simile, or metaphor, but a statement that places speaker-listener—or "addressee"—in relation to one another as tuned instruments. This is the basic hypothesis of Noam Chomsky's work as well and seems to pay no heed whatsoever to the meaning of speaker–listener as two instances in the life history of two organisms. In fact, none of the writers who take this mechanistic approach in methodology really believes in its reality. Jakobson and Halle (1955) write:

> The code of features used by the listener does not exhaust the information he receives from the sounds of the incoming message. From its sound shape he extracts clues to the identity of the sender.

We need only examine the information content—the entropy, if you wish—of the "identity of the sender" to recognize that it is fantastic, something like the ratio of dot–dash in telegraphy to the number of black-white dots at any moment on a television screen. But a dog can look at a television screen and see nothing, as a savage or a child can look into a mirror and, at first, see nothing. Evidently the idea of a code derives from the problem of transmitting verbal messages by nonverbal means. One may say that Michael Ventris and John Chadwick (1958) decoded Linear B by a "dextrous manipulation of the message" but only as a result of a fair knowledge of both the language and the culture involved.

Another source of confusion of mind and brain in linguistic theory, in addition to this great reliance on communication engineering, derives from neurology and neurophysiology. The study of aphasia, to which Freud made a classical contribution before he ever developed the psychoanalytical method, is a rich source of ideas about language function. But it is also a source of serious pitfalls. First of all, despite the capacity of able observers to classify aphasias, every case is different to a significant degree, the more so when the damage is cortical rather than in deeper, pathway structures. The patterns of aphasic difficulties stand in relation to the language symptoms of mental illness as those of organic paralyses or anaesthesias do in relation to hysterical ones. We can distinguish by pattern in a manner that is comparable to the way in which the aeronautical "black box" technique distinguishes between the pattern of a machine failure and an error in human judgment in an air crash. The loss of language in regressive mental illness such as catatonic schizophrenia has no resemblance in pattern to aphasia, nor does the mutism of the autistic child resemble the so-called pattern of mutism in the mental defective.

This differentiation between man and machine, between mind and brain, is essential for further discussion. Clearly machines do not—and never will—"talk," any more than planes will "fly." Birds fly; they live in air. Machines can only propel themselves through the air from place to place. The thing that amuses us so much when we see a swan run splashing and flapping across the water is that his ungainliness is like an airplane, although we forget this clumsiness when we watch a jet liner make its run.

Humans live linguistically. It is essential to their humanity. "Speech is the best show man puts on," Benjamin Lee Whorf (1942) writes. Probably Susanne Langer's (1942) grasp of the peculiar concatenation of social impulse, lalling impulse, and symbolization impulse which drives the child to become "speaking," comes closer to the psychoanalytical theory. But, as we shall see, the key concept, identification, is still missing. Surely the most brilliant work in linguistics with the struggle to cross from brain to mind as its central preoccupation is that of Noam Chomsky. With a rich philological, semantic, and phonological background, and under the influence of communication engineering, he has set out to utilize the philosophical direction set by Russell, Wittgenstein, and Carnap to investigate syntax by divorcing its formal qualities from the semantic aspects (Chomsky, 1957). He does this by constructing a generative grammar composed of the rules of transformation of strings of morphemes. The near-impossible aim is to invent a universal grammar that would differentiate the grammatical from the agrammatic in any language, quite regardless of whether the sentence in question was meaningful, meaningless, or even unthinkable. Of course, he is faced with the task of closing the gap exposed by intuition in the initial stages of his theorizing, for he acknowledges that the division grammatical-agrammatical is in the first instance an intuitive judgment. He also acknowledges that a distinction between surface and depth must be made in grammar, as in meaning, as described by Wittgenstein (1953). But nonetheless he pushes on to insist that in learning language a child must invent grammar before he can comprehend what is being said (Chomsky, 1965). We know that a child, like Koehler's apes, could hardly "invent" joining two sticks together to reach the bananas without being taught. The question, from Augustine to Wittgenstein, has been the nature of the "teaching." It seems extraordinary that a man who can see with such ease the dehumanized aspects of the behaviorist theory of language (Chomsky, 1959) and can comprehend that science aims primarily at insight and not at precision for its own sake should also lose touch with the distinction of man from machine. But what he is working toward is the construction and programming of a translating machine.

In what follows, the concept of *intuition* will be given finer definition and the concept of *learning* will be firmly linked to that of *teaching* to define the context in which speech develops in the child. But first we must consider the other side of the spectrum in linguistic theory where intuition is not eschewed but embraced as a cosmic mysticism which needs to be distinguished from the mystical element in psychoanalysis.

The Mystical Aspects of the Problem

There is an evasion of the problem of the individual mind and its extraordinary development from birth onward which is common to religion in the past—that is, to theological psychology—and anthropology in the present. Jerrold Katz (1964), in fact, calls it a "theological type of mentalism" in dissociating his own work and that of Chomsky from the suspicion of dealing with *mind* as synonymous with *soul* or *spirit*. Clearly their conception of mind is far too neurophysiological for any such link and can barely be dissociated from the mindlessness of behaviorism which knows nothing of mind but only of the "summation of behavior." But psychoanalysis has a very distinct link with theology, both in its essentially introspective method, and in its findings. While all theologies find godhead external to the individual, psychoanalysis finds it internal and here differs from the psychology of Jung or the mysticism of Ouspensky (1949). The form which this cosmic mysticism takes in linguistic science comes from the anthropological and is exemplified in the work of Korzybski. The central idea is that the culture, through its language, imposes limitations on the modes of thought of the individual, thereby attributing to language and culture a reality and continuity which is primary rather than secondary in respect to the individual. In a sense it places culture in relation to the individual as Mendelian-Darwinian theory places the species in relation to the individual member.

On the contrary, everything indicates that heredity in the realm of mind is thoroughly Lamarckian—that is, derived from the acquired characteristics of parts—in transmission from generation to generation (Freud, 1938, S.E. XXII) both in form and content, or, in the case of language, both in syntax and semantics. A certain anticivilization spirit intrudes upon that type of anthropology which sees a superior virtue in the primitive which o'erleaps itself in opposing what it thinks is the prevalent attitude of contempt. But they are wrong to think that it is prevalent amongst scientific workers in this field to consider so-called primitive, or even aboriginal, groups as inferior in intelligence, any more than one thinks of the child in this respect. Whorf (1956) admires the Hopi Indian's model of the universe and considers his tenseless language quite satisfactory because of his unification of space–time:

> The Hopi language is capable of accounting for and describing correctly, in a pragmatic or operational sense, all observable phenomena of the universe. Hence I find it gratuitous to assume that Hopi thinking contains any such notion as the supposed intuitively felt flowing of "time," or that the intuition of a Hopi gives him this as one of its data. Just as it is possible to have any number of geometries other than the Euclidean, which give an equally perfect account of space configuration, so it is possible to have descriptions of the universe, all equally valid, that do not contain our familiar contrasts of time and space. The relativity viewpoint of modern physics is one such view, conceived in mathematical terms, and the Hopi Weltanschaung is another and quite different one, non-mathematical and linguistic.

Of course, the whole problem resides in the idea of "capable of accounting for and describing. . . ." It is a kind of anthropophilia which wishes to see special virtue in the primitive and generously projects into the savage mind, much as people project into children, that peculiar and incompatible mixture of innocence and creativity by which their idealization is consummated. It stems from a distrust of the concepts of development, of individuality and of the adult-infantile differentiation as a qualitative one. We will come later to this point, of distinguishing between child–adult as a descriptive antithesis and infantile-adult as a meta-psychological one.

The mystical element in psychoanalytic theory, on the other hand, resides not in its modes of thought but in the facts of mental life which it appears to discover. It takes the concept *creative* to mean more than what Chomsky (1965) means when he writes:

> In fact, a real understanding of how a language can (in Humboldt's words) "make infinite use of finite means" has developed only within the last thirty years, in the course of studies in the foundations of mathematics. Now that these insights are readily available it is possible to return to the problems that were raised, but not solved, in traditional linguistic theory; and ‚to attempt an explicit formulation of the 'creative' process of language.

By "creative," however, psychoanalysts mean something more like "raising to a new level of self-perpetuating orderliness." Chomsky may stand in awe of infinite quantity, of "generative power," but we stand in awe of qualitative transmutation. He speaks of the "power" of ideas and methods in the mathematical sense; we must speak of the "penetration" of "insights," meaning the ever greater and more detailed location of the mystery, not its solution. In a sense this falls foul of the modern spirit in science, which Medawar (1970), for instance, celebrates in honor of the discovery of the double-helix structure in the chemistry of genes. But it is naive—or is it arrogant—to think that exposition of mechanism answers the full range of meanings in the question "How?"—setting aside the teleological "Why?" This turning to mathematics grows out of a certain despair about "intuition" as beyond investigation, not because of the solipsistic element in individual minds, but because of their disbelief in the possibility of a method to "look into the fluent speaker's head" (Katz, 1964) without "building a fire in a wooden stove." (Twaddell, 1935) We claim, of course, that the psychoanalytical method of Freud does exactly this, but "into the mind," not "the head."

The nature of the cosmic mysticism contained in such a theory of the relativity of language may not seem one that impinges as a serious problem in one's thinking, until its implications for epistemological theory are recognized. The theory of knowledge implied in Whorf's book gives con-

creteness to words as containing meaning in themselves, as full and therefore available for exploration. It indicates that all knowledge *exists* and
awaits discovery. The mind of God manifests itself in the word.

Psychoanalysis also has a mystical element which relates to epistemology, but it also places the scene of transactions differently and sees
the relation to language in a more creative sense. Thus we may recognize
the primal combined object, breast and nipple, as truly the source of
knowledge, in that thinking is an unconscious mental activity the scene of
which is the baby–breast relationship: That is, internal *teaching* means
that the breast *knows everything,* in a categorical sense. It is omniscient,
contains all knowledge—not of course in terms of external reality but as
category of meaning in psychic reality. Upon this stage, the words, given
as empty containers by external objects, are filled with meaning by the
internal breast. But this is a lifetime process, by which experiences (Bion,
1962) may be assimilated to fill with meaning the verbal categories in
ever-expanding levels of abstraction. The filling of old words with new
meaning need never destroy, nor even overwhelm or obscure their old
meaning, so marvellously contextual in actual language usage is the selection of the particular aspect of a word's multitudinous contents. More suspect as a process is the invention of new words—or at least new morphemes. The taxonomic aspect of scientific investigation has always
avoided this problem by the appropriation of Greek and Latin morphemes for compounding into names for objects or processes of enquiry.
In a prescientific era in any field, before "facts" have been delineated, a
phenomenon described by someone tends to have his name stuck to it as
an identification tag, like Spoonerism or Bright's disease. These are not
really new morphemes, but, rather like mathematicians' X and Y, identification tags for unknowns.

One would expect, therefore, that the lexical evolution of language
would involve an *apparent* continual expansion and a *real* continual sim-
plification, at the morpheme level. In this way the language of science
describing the outer world and the language of poetry describing the inner
world can be seen to follow the same basis of development. We will claim
that psychoanalysis bridges this gap for the first time in the unique direction, a scientific poetry. Poetic science, on the other hand, is as old as
religion.

THE RELATION OF WORDS, LANGUAGE, AND IMAGE

Having examined some of the work in modern linguistic theory and
its relation to psychoanalysis in both methodology and basic premises, we

may move on to examine the relation of language to image as seen in the
analysis of dreams. It is hoped that this will enable us to formulate a truly
psychoanalytical conception of language function which can then be used
to examine and formulate the problem of mutism in the autistic child and
the catatonic schizophrenic.

I will present clinical material from two patients to illustrate two different aspects of the problem. First, material of a gifted neurotic patient
will be used to show how deep are the roots of language in unconscious
plastic phantasy. Second, material of a chronic psychotic, also a gifted
person nonverbally, will be brought to show the fundamentally nonverbal
nature of vocalization and its relation to infantile babbling.

First, then, the clinical material. A young poet, under the influence
of recent high praise of his latest volume, shortly after the Christmas
break and approaching the weekend, brought this dream.

> He was going to the home of Elizabeth Taylor and Richard Burton
> to borrow a car, which, in the dream, was in fact the car he owns. It
> seemed necessary to go from the gravelled drive, through the house
> into the garden to reach the rear where the car was parked. There
> was a huge party going on, in the house and garden, and, since he
> had not been invited, he pushed his way through the crowd averting
> his gaze so as not to be stopped in conversation by people he knew,
> lest he be thought by the Burtons to be gate-crashing. In the gar-
> den a commotion was going on around a little pavilion which was a
> first aid station, for a young woman named Miss Spoonerism had
> died and was being carried out on a stretcher. When a woman called
> to him, he greeted her as "Elizabeth" but she said she was not Eliz-
> abeth, and he realized that it was the secretary or someone whom
> he seemed to know, and he told her a lie, that he had come to return
> Elizabeth's car, and pressed on.

His associations were that he and his wife were invited to lunch that
day by the ex-wife, also named Elizabeth, of a movie star, also named
Richard. But he had decided not to go. He could not remember exactly
what a spoonerism was and had looked it up, as he had once before and
found the example "crushing blow–blushing crow." In the dream there
was nothing strange in the name Miss Spoonerism, no joke or unreality.

The essential elements I wish to emphasize are: the passage through
the front to rear, the intrusion, the misidentification, the lying reversal,
the death, the associated equivocation regarding the marital status of the
two Elizabeth–Richard couples, the equivocation about his welcome
(avoiding greetings yet fearful of being accused of gate-crashing).

The analytic background of the dream is very important. This patient had lost his father at an early age and his mother had never remarried, though much sought after by men and passionate in her nature. The patient was married to a charming, girlish woman who was experienced by him, as documented by innumerable dreams, as a part of his mother, namely the breast–buttock with which he had an erotic and possessive relationship. Since he allowed no father in his inner world but presided there himself as the little husband, the differentiation of levels, adult-infantile, was extremely confused. He could not, therefore, in his infantile omnipotence recognize that he treated his women badly from a manly point of view. This was repeated in the maternal transference in the analysis, in which his love affair with the analytic method produced the feeling of being the ideal patient, so that he could never recognize how he controlled his material, denigrated the analyst as the father, and begrudged respect and gratitude to both father and mother aspects of the analyst, represented specifically in insisting that he could only pay a low fee as his future income was uncertain.

The prolonged struggle over these issues had been broken momentarily just before the holiday when he had volunteered to pay the regular fee under the pressure of depressive anxiety about the analyst's safety, stirred by a very disturbing dream in *which his mother, hand to her breast, complained of his selfishness, that after a heavy meal he just goes off to a party with his wife, leaving her to worry about their baby.* But the differentiation and insight were lost almost immediately after the holiday break and the old struggle resumed. Thus the dreamer came under the pressure of having received a bill which he was neglecting to pay without reason and even neglecting to mention.

During the session of the "Miss Spoonerism" dream the patient was very resistant to the analysis but came to the next session in a very different mood, paid his bill, and brought material which had been withheld. What he had neglected to reveal in the previous session was that he was planning that night to see a new film with the Burtons in which Elizabeth played the part of a dying woman with no husband.

The analysis of the dream was, in substance, as follows. In it we saw once again the intrusive little boy (the gate-crasher) who could not allow Mummy (Elizabeth) to be married to Daddy (Richard—the divorced one whose wife's invitation to lunch he planned to refuse) because he could not resist borrowing and owning the breast (his white car) which he thought he reached by going inside the mother's genital in intercourse (the party), since he was so confused between breast and buttock (front and back garden). In this confusion he could not comprehend the idea of being unwelcome to Mummy's genital when he was invited to her breast (luncheon invitation by the other Elizabeth), but he avoided noticing the signs of being unwelcome (averting his gaze as he pushed through) and lied about his intentions (to return rather than borrow the car). Only when we link the dream to that before the holiday (his mother with hand-to-breast complaining of his selfishness) and the film he was planning to see that following evening (of the dying divorcée) can we understand that the seemingly irrelevant death of Miss Spoonerism means that the Mummy who treats the little boy's intrusiveness and his confusing front (breast and genital) with back (buttock and rectum) as a joke (blushing crow) is a dying Mummy (crushing blow) due to lack of love (the dying divorcée, visited in the film only by a gigolo, the Angel of Death). The dream of a poet, gifted to a high degree in both visual and verbal representation.

In the case of our dreamer, we have already formulated his material insofar as it refers to the structure of the transference and his internal situation, with its genital references. Let us turn now to its linguistic reference. Our poet's reliance upon language as an omnipotent tool can be traced to a very early time in his life. It was a matter of family folklore how at the age of six he had routed, with a stream of logic and invective, some political police of his native country who had come to investigate his father's absence. He was also proud of the speed and totality with which he had changed from his native tongue to English in prepubertal years. His verbal gifts took him into the world at an early age, cutting short his formal education. His facility with language enabled him to converse freely with educated people in a wide variety of fields to a degree which hid not only from others but from himself the shallowness of his understanding of science, history, philosophy, mathematics, economics, and politics. His wit was equally reliable as a weapon against persecutors and as a tool of seduction.

Similarly, his contact with unconscious phantasy was unusually detailed and consistent. Seldom did he come to a session without a dream, vividly recalled and ably communicated in both its visual and affective content.

The linguistic elements manifest in the dream are the following:

1. Equation of objects through equation of their names—the two Elizabeth–Richard couples. By this equation he was able to uncouple them at the verbal level.
2. The many levels at which the reversal of meaning is worked over in the dream and associations (going from the front garden to the rear garage; the reversal of borrow-return; the balance of Elizabeth–not-Elizabeth; the uncertainty of the guest-gate-crasher; the contrast of party–death).

3. This last contrast also highlights the manner in which affects are reversed and is the key to the defensive use of the spoonerism as a form of humor, as in crushing blow–blushing crow. Thus we are dealing with a manic defense against depressive pain, the items of which can be listed from the dream as follows:

The divorcing of the Elizabeth–Richard couple
The gate-crashing
The borrowing of the car, presumably without permission

All these acts of selfishness, which cause the mother to put her hand to her breast and threaten the crushing blow of her death, are reversed by the spoonerism joke, of his manic reparative lie that he is returning the car. The death of Miss Spoonerism, therefore, means that he must choose between the death of his mother, Mrs., and the death of his unmarried wife-breast, Miss S., who thinks little boys' verbal tricks are so funny that their intrusiveness is altered by the death of Miss Spoonerism, from the white car he came to borrow to the first-aid pavilion, thereby also differentiating breast from buttock (the white car in the back of the genital house-garden-party).

4. The syntactic implication—or can one boldly say structure—of the dream is of special interest. Coming as the dream does at a point in the analysis when insight has begun to alter the economic balance between paranoid-schizoid and depressive (PS ↔ D) relations to internal objects and in the transference, it can be seen to have a structure which I have come to call the *T-junction* structure after its most unequivocal type of representation in dreams. Its phrase structure can be seen as, "I used to . . . , but since . . . , I now. . . ." Namely: "I used to feel free to intrude into Mummy's and Daddy's sexual relationship, borrowing Mummy's bottom for my pleasure, concerned only that other people should not think that I did so without Mummy's permission; but since the death of her sense of humor about my selfishness, I now realize that it is a lie when I say that by defecating I return something borrowed from her, and anyhow I realize that such intimacies have only been allowed by the maid and not by Mummy at all."

The thesis is that understanding and speaking *language* comes to children before the use of *words*, that language function is a deeply unconscious process and not, as Freud was inclined to think, one which takes place between conscious and preconscious levels, topographically speaking, as a means of anchoring thought in consciousness. In distinguishing between the use of language and the use of words, we imply that it is this transition which has a certain topographical significance, not the presumed process of changing image to language.

Another way of stating the same idea would be to distinguish between the use of language as a mode of operation of projective identification—that is, for the communication of states of mind—while words are used for the transmission of information from mind to mind. The former involves a degree of regression to narcissism in that object–self boundaries are in some measure surrendered for the moment.

This does not, however, correspond to Piaget's "egocentric speech" of children, which is really vocalized inner speech. It may be useful, for instance, to distinguish between two types of unintelligibility. Thus egocentric speech in its early stages, corresponds to infantile "babbling," in which the child expects to be as well understood by its external object as it apparently is by its internal one. In contrast are those mistakes based on homonymity, inaccurate reproduction of morphemes, reversals of phonemes of the spoonerism type, substitution of antonyms, unconscious double-entendres, dangling particles, and the like (Empson, 1930). These comprise children's "howlers" which so entertain the adult world. In the "Miss Spoonerism" dream I have illustrated the way in which the unconscious plastic phantasy and the verbal manipulation are linked. In that instance the material came from a highly gifted and not-very-ill adult. When we turn to psychotic patients, we find evidence of the first type of difficulty, of babbling language, in which again we find a certain confusion, but this time of phonemes and underlying thought. It produces a type of humor also of the "drunken" type often employed by comics. Let me give an example:

A woman in her mid-thirties, but still looking like a frail, pretty, pubertal child, had been in hospital for eight years, variously considered manic-depressive or catatonic at different times. Her life on the ward was divided between lifeless periods in bed and driven activity as scullery maid and general dogsbody under the tyrannical control of another chronic patient, Millie, who had something of a coterie. With this clique Millie seemed to my patient to dominate the ward and intimidate the staff. The peace was only occasionally broken by someone—my patient included—"going up the wall" or "smashing." The former consisted of screaming assertiveness and the latter of breaking dishes and crockery. These outbreaks were attributed to various intrusions from staff or visitors into Millie's Pax Romana: "If only they would let us alone," she often said. The analysis, to which the patient was being brought by car and nurse, was limited by her to two sessions per week as the maximum intrusion into this status quo that she could tolerate.

As the Easter break in analysis approached, she began "smashing" and "going up the wall" and attempting suicide on returning to the hospital after each session. But she struggled against these states, had herself put into a seclusion room, and cooperated very well in the sessions. To the penultimate session she brought the following dreams:

A. *Millie was cutting up and handing about lettuce.*

B. *Amy might smash a little tank outside her room.*

Associations, sometimes stimulated by the analyst's enquiring noises or repetition of words from the dream, brought further information. Sometimes they have lettuce with the meals, but there was no meal with the dream. Amy is a "smasher" and gets upset at holidays when the ward empties of its less permanent residents as relatives take the less ill home for visiting. The "tank" was of glass, big enough to hold about a pint, with gradations like a thermometer. The dreams were interpreted as meaning: If you would let us alone (lettuce alone) and not stir feelings of love (Amy), we would not destroy our capacity (holding about a pint) for gratitude (the tank) when left alone (Easter break).

It is characteristic of the patient, as she totters from the room at the end of the session, to mumble tearfully at the door, either "m'soy" (I'm sorry) if she has brought no dreams or "Thany" (Thank you) if she has. She weeps on the trip back to the hospital and is filled by suicidal impulses to throw herself from the taxi, to strangle herself with her scarf, to poison herself with secreted pills.

One can clearly see the image, as of concentration camp children, huddled together in utter aversion to the brutality of the adult world, scavenging for food and sharing it out, their slogan "let us alone," "lettuce alone," and "ledusalown" in ever louder, more defiant, more dysarthric chant—"LESALON."

Outside the door of the consulting room is the mad-house of the adult world where she had drifted in confusion from bed to bed during the years before hospitalization, searching for a love-object, that is, an object to fill her with love. But it would have to be an interminable and uninterrupted process, for the removal of the nipple from her mouth (the tank–baby bottle outside Amy's door) confronts her with the facts of having emptied her object. This fact renders her gratitude so painful that she smashes the object, her perception of the sacrifice for her, her capacity to remember the feed, her realization of a good object. "Let us alone." "Say thank you." "Lettuce alone." Say "thanks." LESALON!!

By extrapolating the phonemic implication of the two dreams within the social situation to which they refer—namely, the ward of the mental hospital on the one hand and the analytic situation on the other—I have hoped to demonstrate the language deterioration which is a marked trend—variable, but marked—in this patient. Not only does the morphemic structure seem to melt away into confused homonymity, but the concreteness of the image ("lettuce alone" and "tank") seems to enter into an oscillating, echoing relation to the phonemic structure. The extrapolation to drunken babbling is, I think, an unmistakeable trend. The main point is that the "language" is nonetheless preserved, despite the flux of morphemic, phonemic, and syntactic structures. In the two images, Millie passing out the lettuce and Amy with the tank outside her door, the language of "Let us alone" and "Thank you" is preserved in the childish sense in which inner speech develops, in ignorance of words and grammar, as statements of states of mind, namely *withdrawal* and *gratitude* respectively.

Language, then, we are suggesting, is primarily a function of unconscious phantasy which employs projective identification as its mode of communication. The substance of its communications is states of mind. Its means of communication are fundamentally primitive, namely, song and dance. Its motive is the communication of states of mind; its information content relates primarily to psychic reality and thus to the realm of experience relevant to art, religion, courting, and combat. The subtlety of its content as regards range and intensity of emotion, complexity, levels of abstraction, and logical operations is such as can only be approximated verbally by the poet. Its history must, of course, have antedated verbal language by innumerable millenia and have reached the level of development which we experience at a time when communication of information about external reality was still limited to pointing. This same sequence is, of course, repeated in childhood development, where the elaborate communication between mother and child, consisting of sound and gesture approximating to song and dance, stands in marked contrast to the difficulty of pointing-and-naming in regard to the facts of external reality.

We are thus suggesting a two-step theory of language development: a first step consisting of the realization by the child of its instinctual capacity for inner language, for the internal and external publication (see Bion, 1962) of states of mind; and a second step consisting of the adaptation of this language to the description of external reality by means of verbalization, meaning the delineation of morphemes within the "strings" (see Chomsky, 1965) of phonemes. In this conception, grammar or syntax is seen as a function of inner language. Hence its delineation is necessarily intuitive, placing the grammarian in the position of an after-the-fact judgment of the grammatic-agrammatic differentiation which has no more validity as a value judgment than to say that French is inferior to English. This, I believe, is the point that Whorf would like to make but, lacking the conceptual equipment to distinguish between internal and external language, miscarries onto pleading the moral equality of primitive language, which is irrelevant. Chomsky, on the other hand, being bound to an information-theory conception of language, conceives of grammar as conventional, a carrier for bits of meaning which can be introduced into the empty containers of the carrier in infinite variation, some sensical and others nonsensical.

In our theory, grammar would stand in an absolutely bound relation to the language of unconscious phantasy in something of the same relation as a scale of tones stands to a body of music, or as the particular set of axioms stands in relation to the body of a particular geometry, or as a particular set of "natural rights" stands in relation to a body of law and what the "courts will do in fact." (Lerner, 1943).

In this sense, Chomsky is right to think that grammar "generates" language, but not because of the existence of a set of "rules" separate from meaning. Rather, it is the set of basic meanings in relation to time, space, person, and logical operations which determines the transformation of inner language into inner speech through verbalization. In a sense, Freud was right to think that words had a special relation to consciousness, in that attention, which is the special province of consciousness, is directed by verbalization into items of perception which would otherwise command no attention, just as movement of an item in a static visual field immediately commands attention.

Ernst Cassirer (1923) has treated langauge as one of the many possible symbolic forms by which cognition may objectify itself through and in the action of the mind. In this way he has applied Kant's principles of form in epistemology so as to get beyond the usual philosophical bias of considering the word and the idea to be identical. I say *bias,* for philosophers are "verbal" people as artists are "visual" and musicians "auditory" in their spontaneous preferred mode of representation. It is a view to which I believe Chomsky (1966) also has come more recently. It is a view which is implicit in the psychoanalytical theory of the mind, even though Freud himself equivocated to some extent. At least one cannot really tell how his earlier topographical view of the stratification of the mind in terms of levels of consciousness and his later structural concept fuse with one another in regard to the relationship of verbalization to thought. There is very little to be found outside the *Project* (1897) and Chapter VII of *The Interpretation of Dreams* (1900) which are both, in a sense, preanalytic from the point of view of clinical method of enquiry. My impression is that word and symbol remained very closely bound in his mind as representation of meaning in a far more rigid sense than Wittgenstein's "seeing as" as a basis of thought, and in a far more restricted sense in regard to the meaning of words than Russell's meta-levels. The

concept of perception as creative-active process did not really enter psychoanalytical thought until the work of Paul Schilder(1942). Freud's concept of perceptual consciousness was, for instance, a far more photo-
graphic, "copy" theory.

How far have we come, then, in our argument? In effect I have
brought some clinical material, and especially the dreams, of a neurotic
poet and a psychotic painter, to broach the thesis that language, in its
truest meaning, is a process that emerges from unconscious phantasy and
that formal representations of various sorts organize these phantasies in
the publishable forms which can serve as modes of communication of
mental states. Language is one of these several forms. Any formed representation may be built secondarily into a notational system for communicating information about external reality. This adaptation is accomplished by a kind of ellipsis which omits mention of the cognitive process,
as, for instance, in Russell's (1940) "I am cat-perceptive" as the correct
semantic form for "It is a cat." A tertiary system may represent this in
turn, as in written language or musical notation. $1°, 2°, 3° \ldots$ etc., systems
of notation are quite independent of $1°, 2°, 3° \ldots$ etc., meta-levels of
meaning.

Now the question arises as to whether words are a system of notation
by means of which people exchange information about language as a phenomenon or object of cognition, in the outside world; or do they, in themselves, comprise a symbolic form, by means of which cognition is represented in the mind? Language we accept as a symbolic form. But words?
Morphemes? Phonemes? Letters? Ideograms? Hieroglyphics? What I am
suggesting is that we consider *vocalization* as the symbolic form and *verbalization* as its corresponding notational system.

But where does that leave us in regard to grammar? If we return to
our two dreamers, we may most cogently express the answer to this question in terms of the linguistic analyses of two sets of dreams. In the case
of our poet, the theme of the dream over all is that something used to
amuse mother and no longer does, and the patient has himself begun to
understand that it is not funny in the sense of witty-amusing but only in
the sense of triumphant-amusing. The joke at the level of the language of
unconscious phantasy involves a little boy who is such an amusing gate-
crasher at Mummy's bottom that she does not really need any other sexual partner. But Mummy-with-her-hand-to-her-heart has put a stop to
his joke. The symbolic form at this level is visual phantasy.

But the dream hints that the same joke has a representation in
another symbolic form, vocalization, called technically a spoonerism.
Associations of the patient indicate this to be the juxtaposition of crushing blow and blushing crow. One would be inclined to think that jokes of

this sort were the dysarthric, "drunken" type of humor so popular with children and leave the matter there, as if it were explained. It would be just as easy to say that the custard-pie-in-the-face routine "just is" funny.

Freud's (1938, S.E. VIII) approach to wit goes some distance to account for the emotionality tapped by jokes, but it does very little to explain why the joke is funny or to investigate different categories of humor. Again this work was preanalytic from the methodological point of view. Nothing but the method of clinical psychoanalysis could plumb the depths of the humor in a Spoonerism and demonstrate its immediate source of comic effect in the preverbal roguishness of a little boy.

In other words, I am suggesting that in the case of the spoonerism the humor comes from below the level of verbalization and has nothing fundamentally to do with words or their manipulation, but instead with *meaning* and its manipulation. However, in its vocalized form the humor had adopted a verbal context for its expression. Now had the dream of our psychotic painter been instead a charade acted out as a game at a party, the humor would have been, presumably, a pun on the homonymous words "Let us alone" and "Lettuce alone." It was not a joke, but rather demonstrated that the patient's regression in her dreams and in the transference had proceeded to a preverbal type of vocalization, characteristic of preschool childhood, in which homonymity of sound and identity of meaning were not distinguished. We might conclude, then, that the hypothetical charade at the party would involve a bit of humor the unconscious determination of which involved some juxtaposition of adult and infantile mentality in which confusion at infantile level was held up to ridicule either cruelly as older children may with younger ones, or gently as adults may with young children.

We seem, therefore, to have elaborated a two-tier theory of speech: that it consists of a system of vocalization as the publishable symbolic form of one current of unconscious phantasy and therefore of thought; and that this system lends itself to application, as verbalization, to a notational system for the communication of information about the outside world. Accordingly, we think that grammar also is two-tiered. Depth (unconscious) grammar includes the phonemic-morphemic elements of vocalization in all its musical aspects (including the postural and mimed aspects related to dance and dramatization) as well as the logical operations of syntax which are implied in the juxtapositions contained in the unconscious phantasy sequence. Surface grammar contains all those modifications of vocalizations which the communication of information about the external world requires in order to minimize the many possible forms of ambiguity—and therefore confusion.

Colliqual speech is very poorly equipped for this latter task and is notoriously "agrammatical" in the sense of surface grammar. But, correspondingly, speech which *is* grammatically correct in this surface sense is notoriously poor for the communication of states of mind. Dare we suggest that the technical skill of the poet resides exactly here, in the bringing together of deep and surface grammars?

In a peculiar, split sort of way, this is precisely the view to which Chomsky (1966) appears to arrive when he reviews the *history* of linguistic thought (from Descartes, throught the Port Royal grammarians, to von Humboldt, Leibniz, and modern philosophy). How he manages to harmonize it, nonetheless, with his conception of a universal generative grammar remains a mystery to me.

LANGUAGE PATHOLOGY IN SCHIZOPHRENIA AND AUTISM

I will refer first of all to clinical experience in the realms of language function from the analyses of two adolescent chronic schizophrenic patients, both hospitalized and brought to analysis at the analyst's consulting rooms, one a 16-year-old girl, Pauline, the other a 21-year-old "boy," Christopher.

Pauline was a highly intelligent and verbally gifted child whose illness commenced with a full-blown delusional system that appeared one morning following a dream. But this florid illness, hebephrenic in quality and paranoid in content, was the sequel to a severe psychotic depressive phase that had appeared in puberty. I will not trace in detail her delusional system here. Suffice it to say that she was the captive of a rich man who had bought her from her parents as the subject for a huge research project in schizophrenia, for which purpose she was confined to a movie set where nothing was real—neither the air nor the sets nor the actors—nothing but herself. As everything was being followed by television cameras, cleverly hidden, Pauline's every utterance and gesture was studied, theatrical, controlled. However, the theme of control revealed a sharp split in her thinking. Her own behavior was experienced as being controlled by the rich man and the circumstances of the experiment, so that no sense of personal responsibility attached to her words or behavior. In sharp contrast to this, her relationship to the analyst, who seemed mysteriously to have the same name as the "rich man," was one of omnipotent control over his words and behavior. It was necessary for the analyst to curtail every extraneous movement or facial expression, since anything of the sort resulted in an outbreak of manic triumph, mockery, and contempt. With flawless logic she would explain, "You can't seem to control yourself. However, as there are only two of us here, I must be controlling you."

While these manic outbursts initially followed, as I have said, upon the analyst's extraneous movements, such as crossing a leg or scratching an itch, they gradually spread to the analytical activity itself. The effect was intimidating indeed, so that it became noticeable that an inward struggle was necessary to overcome the inertia of silence. But the consequence of the struggle to maintain the interpretive activity was a most undesirable one indeed, from the scientific viewpoint if not necessarily from the therapeutic one. The interpretive perseverance appeared to have the effect of inhibiting the patient's impulse to speak, so that her performance gradually subsided into mime. But it became noticeable as well that she seemed to look at the analyst less and less often after the first few moments of the session, but rather directed her attention outside the window. From the behavior of her eyes, which now for months commenced a most complicated and bizarre sequence of blinking and staring, it was possible to construe that she was using her eyes as a camera and projecting an image of the analyst outside the room in order to carry on her controlling relationship with a more suitably docile subject. When this was finally pieced together and presented as an organized interpretation, it drew an immediate angry response, a veritable bonanza of confirmation in this desert of negativism: "Pictures are just as good as people."

Three years of treatment had resulted in a major shift in the patient's pathology but can hardly be held as a therapeutic triumph. From being a verbose and theatrical actress in a delusional paranoid system, Pauline had become progressively catatonic. From being constantly before the camera, she had metamorphosed into the camera itself. From being held under minute control by her persecutor, the rich man, she had transformed into the persecutor who, with her camera eyes, held the picture of Dr. Meltzer, which was "just as good" as the person, under tyrannical control. In the process her impulse to verbal behavior dried up. A sad tribute to the power of language, one must admit!

At the time that Christopher came to treatment he had already been in a catatonic state of deepening intensity for six years, having drifted into it relentlessly following several years of increasing confusion, paranoid anxieties, and uncontrollable behavior commencing with puberty. By the age of 21 he looked like a dishevelled child of 12, or more like a sad little clown or even a rag doll, his verbal responses being largely restricted to "yeh, yeh" and "dunno, dunno." At moments of rage he would scream at his hallucinated persecutor such phrases as "Get off my back," or "Stop fucking me," alternating with expressions of wilful negativism, "I'm going to have my way," or "I'm going to do as I please." His voice was toneless, unrhythmic, mechanical. He wet and soiled constantly, masturbated, tore off his clothes, slammed doors, and giggled. Alternatively, he could sit

motionless for the entire session. One need not fear exaggeration in estimating that he had struck bottom in his regression. It is of some interest
to note the similarity to the "Let us alone" of the manic-depressive
woman. During analysis she developed a very strong maternal transference to the nurse who brought her to treatment, underwent a severe
regression to infantile behavior including loss of speech and incontinence.
When the analysis was broken off at the insistence of the hospital, she
promptly recovered, but not to her previous behavior. Rather, she manifested a remission of her psychosis which lasted many months, but was
accompanied by a firm refusal of further analytical treatment.

The difference, however, between the language loss in the two types
of patient is of great importance. The regression to infantile behavior and
preverbal vocalization in the manic-depressive woman was not accompanied by any loss of capacity nor drive to communicate her states of mind,
in which every primitive device of mime, vocalization, and facial expression seemed rather to be heightened in compensation for the loss of verbal
capacity. In contrast, Pauline in her later stages and Christopher early on
in the analysis were both deeply involved in the repetitive behavior which
manifested their omnipotent control over objects. So restricted was their
mental content by this overriding preoccupation that the continuity of
perception–thought–memory appeared to crumble. The consequence was
the appearance in Pauline of near mutism in the sessions, while a converse
course could be traced as Christopher's catatonia lessened. As his
automatisms gave way to contact in the analysis, communication
increased in the sense that there was an increase in the amount of language but it consisted of two types of verbal activity. The first of these
consisted of the mentioning of song titles or film titles which he had
encountered on TV. These utterances were quite smoothly, if mechanically, delivered. But the second type of communication was hesitant, broken, often trailing off into "dunno, dunno." Its content was memories, for
all the world like the archaeologist's broken pot-sherds, the debris of the
"catastrophe," as Bion calls it. I cannot devote space here to tracing the
therapeutic process that induced this return to communication, but I do
wish to indicate that its essence resided in his return from "nowhere"
where he was eternally 18 (his age at entrance to mental hospital) to a
world of time–space consisting of the triangle of his travels between hospital, consulting room, and the home of the nursing couple who kept him
four days of the week, with the relevant time reversals.

Of the two types of verbal activity, the song and film titles served to
communicate his transference experiences, while the broken language
with which he struggled, sometimes piteously, served for an attempt to
reconstruct the past and thereby his identity. This fact showed itself also

in the gradual relinquishment of the name "Ivan" in favor of his proper Christian name, "Christopher." In the course of this return from absolute despair (in the Kierkegaardian sense)—I do not think it correct to consider it as absolute or primary narcissism—the renewed urge to communicate, sporadic to be sure, produced what I would consider to be the two primary forms of infantile linquistic activity, singing and lalling. While Christopher could not sing the song, his indication of the title left little doubt that the internal content was the song itself, that the music as well as the words were definitive of his state of mind in the transference. But the lalling, or playing with vocalization, which gives such evident pleasure to the baby and the small child, was obviously a matter of struggle and torment to Christopher, as his return to language usage was greatly opposed by the destructive part of his personality. This opposition manifested itself in such concrete ways as his hand being put into his mouth, his bottom demanding to be scratched or his needing suddenly to leave the room to "have a wash." The degree of his enslavement to his persecutor could actually be quantified by observing the degree to which his right hand was withdrawn from the outside to the inside of his shirt sleeve. During this time the fact of his attention to and comprehension of the interpretations was indicated by a complicated and interesting piece of treachery. At the end of each session, as he entered the car to return to the hospital or the home of the nursing couple, he would report to the driver a slightly garbled, but effectively distorted, ridiculed, and disguised version of the analyst's last interpretation. It bore a distinct resemblance to the reporting in the session of the song and film titles.

The purpose of this lengthy description is to establish or at least to illustrate the thesis that in the schizophrenic we see a catastrophe which has destroyed the internal basis of identity and carried to destruction with it the basis of language, both as vocalization and as verbalization. In Christopher's moves toward recovery one can discern the outlines of the infantile process of speech development in its two distinct dimensions. One dimension is that of learning by introjection from an internal object (the song titles) by means of which vocal and verbal internal objects, suitable for introjective identification, are developed. The second dimension is that of lalling, playing with the vocal apparatus within the narcissistic organization of the personality. It is through the latter operation that language facility is developed which makes "infinite use of finite means" (Chomsky, 1965) for expressing the particulars of individual experience, the summation of which comprises the true *identity* of a person (as distinguished from his *sense of identity* which is variously compounded).

It will have become apparent that we have by now formulated a more complicated thesis regarding the nature of linguistic capacity: that it is

structured in a two-tier manner, deep vocalization and surface verbalization (deep and surface not having any necessary reference to levels of consciousness, as we are assuming language function to be fundamentally
unconscious in its operation and variably available to conscious scrutiny,
modification, and deployment); that it has developmental roots in two
types of mental processes, one object-related (introjection and introjective
identification, modified variably by projective identification, mimicry, and
the spectrum of narcissistic modes of identification as yet not clearly
defined in psychoanalysis) and the other narcissistic (lalling, ego-centric
speech in the true sense). It is fundamentally different from the current
linguistic theories and probably true to the direction pursued in the
Philosophical Investigations (Wittgenstein, 1953). To give an example,
Roman Jakobson (Jakobson & Halle, 1955, p. 74) writes:

> The gradual regression of the sound pattern in aphasics regularly reverses
> the order of children's phonemic acquisitions. This regression involves an
> inflation of homonyms and a decrease of vocabulary. If this two-fold—pho-
> nemic and lexical—disablement progresses further, the last residues of
> speech are one-phoneme–one-word–one-sentence utterances: the patient
> relapses into the initial phases of infants' linguistic development or even to
> their pre-lingual stage.

Compare the following from Wittgenstein (1953, p. 20):

> Someone who did not understand our language, a foreigner, who had fairly
> often heard someone giving the order: 'Bring me a slab!' might believe this
> whole series of sounds was one word, corresponding perhaps to the word for
> 'building stone' in his language. If he himself had then given this order per-
> haps he would have pronounced it differently, and we should say: he pro-
> nounces it so oddly because he takes it for a single word.

The comical possibilities of this confusion between naming and discourse
are infinite. One famous joke describes the wealthy old Jewish widower
taking his first trip abroad after retiring, invited to the captain's table,
replying to the captain's bow and "Bon appetit" with an identical bow and
"Goldberg," night after night until a fellow passenger takes pity and
explains. The following night he of course preempts the greeting with a
bow and "Bon appetit," to which the Captain, of course, politely replies,
"Goldberg!"

Anyone who has observed infants knows full well that they converse
in discourse, not naming, to their mother's chatter and that when their
mothers begin to "teach" them words, it has the function of clarifying
their discourse, of correcting pronunciation, not of teaching names of
words: "Goldberg," "Bringmeaslab," "dunno," "lesalon." Semantic units
such as these belong to the realm of mind, which Jakobson confuses with
the realm of brain in erudite naiveté. The logic of his argument, trans-

posed for instance to the area of motility, would run something like this: babies cannot walk and have a positive Babinski; victims of bleeding in the pyramidal tracts cannot walk and have a positive Babinski; the pyramidal tracts are not yet myelinized in the newborn human; therefore the stroke causes the victim to "relapse into the initial phases of infants' (motor) development or even to their (premotor) stage." It is simply a semantic confusion to think that saying "walk" to your horse on the long rein is the same instruction as "Walk down to the corner to buy a newspaper" to your child. Wittgenstein (1953, p. 25) puts the point as follows:

> It is sometimes said that animals do not talk because they lack the mental capacity. And this means: 'they do not think, and that is why they do not talk.' But—they simply do not talk. Or to put it better: they do not use language—if we except the most primitive forms of language. Commanding, questioning, recounting, chatting are as much a part of our *natural history* as walking, eating, drinking, playing.

Paralysis from a cerebrovascular accident is not part of our natural history, only the tendency to suffer from it can be so taken. Similarly, we tend to become stuporous (not unconscious) from a blow on the head, but stupor is not a function of mind, but rather a manifestation of mindlessness. It is as categorically separate from sleep as is death.

This preamble must now serve as the background for our formulation of a theory of the nature and origin of mutism in infantile autism. Let us first state the theory of speech as the point of reference for our theory of autistic speechlessness:

> Speech is a *form of life* (Wittgenstein) and one of several *symbolic forms* (Cassirer) in which mental activity manifests itself in humans. Mental activity is essentially *unconscious* (Freud) and variably available to the *organ of consciousness* (Freud) for deployment in the outside world as *intentional* behavior (Anscombe). Inner speech has both narcissistic and object-related forms, the former having its developmental roots in the process of *lalling* (Langer) and the latter in *thinking* (Bion). *Learning* to speak has two aspects: *the acquisition of language* by listening and introjection of speaking internal objects with which identification may be established by various mechanisms and *development of facility* in language usage by lalling or narcissistic language play. Language function is only partially vocal, comprising in vocalization all the elements of song, mime, and dance. Its primary function is dedicated to *communication of states of mind* between self and object and among parts of the self. It is secondarily adapted as *verbalization* for *communication of information* about the outside world (notation). *Knowledge* (or metalanguage, in its verbal symbolic form-Russell and Carnap) is

acquired, in the first instance, by internal objects (Meltzer) and developed by *thinking* (Bion), through the infantile unconscious relationship to the internal mother and her breast. The internal breast has, as its most fundamental meaning, the categorical quality of *omniscience* (Meltzer), that is, it is experienced as containing all possible knowledge. *Grammar* belongs to the realm of vocalization primarily and is only secondarily purified of its ambiguities for the purposes of communication of information. Different language games (Wittgenstein) are not translatable one to the other in the realm of vocalization, only in that of verbalization. (The language game of psychoanalysis is essentially a vocalization, being dedicated to the expression of states of mind.) *Written language* is a notational system and stands in relation to speech in the same inadequate position as written music to the playing of instruments (Langer, Wittgenstein).

We can now turn our attention to the theory of mutism in infantile autism and compare it with mutism in catatonic schizophrenia and psychotic depressive states.

1. The mutism of autistic children has two aspects, absolute and relative, connected with states of autism and with postautistic (or extraautistic) psychosis or immaturity.

    The absolute mutism of the autistic state is fundamental to the mindlessness of the state, which is produced by the extreme dismantling of common sense (Bion). Consequently, vocalization and verbalization are replaced by the making of meaningless noises which may have a random resemblance to language. Since both narcissistic organization and object relations, internal and external, are obviated by the pathological employment of the primitive obsessional mechanism, neither introjective nor lalling aspects of language learning can take place.

    The relative mutism of the extraautistic state in such children is, first, a manifestation of the impeded acquisition of language induced by the autistic states. Second, acquisition is impeded by the operation of obsessional mechanisms utilized under the sway of possessive jealousy. This jealousy manifests itself in a reluctance to relate to objects by modes of behavior employed by real or imagined rivals within and outside the oedipal conflicts. Communication is therefore by preference carried on through omnipotent employment of sounds and gestures. The developmental difficulty in introjecting objects with an effective inner space further interferes with the acquisition internally of speaking objects (cf. Martin in Meltzer, Bremmer, Hoxter, Weddell, & Wittenberg, 1975), while the tendency to solve problems through obsessional experimentation (cf. Wooshie in Meltzer *et al.*, 1975) rather than by learning-from-experience (Bion) further limits the impulse to employ language. To these five items we can add a sixth, namely, that the lalling impulse appears to lose its force progressively from the age of five or so, with the serious consequence that children who have not become speaking before this age are deprived of the internal mechanism for developing facility.

2. The catatonic state produces mutism by two methods, essentially, one being obsessional and the other regressive. The very sadistic type of obsessional control exerted over internal objects imposes a near-lifelessness upon them, of which silence is a necessary component. Furthermore, the abandonment of the dimension of time and space in favor of the negative world of the delusional system results in a loss of identity, in its most fundamental sense, namely the fragmentation of the life-history (cf. Christopher). With the capacity for thinking limited by the lifeless state of the objects, and the capacity for memory destroyed by the fragmentation of the life-history, speech has only the recourse of immediate recall for its employment. The consequence is a parrot-like performance which approximates to echolalia. Even this is curtailed by the command over attention exerted by the persecutor and his delusional system (see Meltzer, 1973).

3. Psychotic depressive states tend to produce mutism by two types of mechanisms, projective identification and regression. The sadistic employment of projective identification, out of motives of envy and jealousy, does severe damage to the internal objects, either in the realm of representation of physical or mental attributes. The capacity of internal objects to speak may be one such attribute that is attacked, with the result that the patient finds himself identified both by introjection and projection with a muted object. Furthermore, regression within the infantile organization may reduce the use of speech to infantile levels of vocalization, even to the level of babbling.

REFERENCES

Bion, W. *Learning from experience.* London: Heinemann, 1962.
Cassirer, E. *The philosophy of symbolic forms, Vol. 1: Language.* New Haven, Conn.: Yale University Press, 1923.
Chadwick, J. *The dicipherment of Linear B.* Cambridge, England: Cambridge University Press, 1958.
Chomsky, N. *Syntactic structures.* The Hague: Mouton, 1957.

Chomsky, N. Review of Skinner's *Verbal behavior*. *Language*, 1959, *35*, 26–58.

Chomsky, N. *Aspects of the theory of syntax*. Cambridge, Mass.: M.I.T. Press, 1965 (see Methodological Preliminaries, pp. 3–62).

Chomsky, N. *Cartesian linguistics*. New York: Harper and Row, 1966.

Empson, W. *Seven types of ambiguity*. London: Chatto and Windus, 1930.

Freud, S. *An outline of psychoanalysis*. London: Hogarth, 1938.

Jakobson, R. & Halle, M. *Fundamentals of language*. The Hague: Mouton, 1955.

Katz, J. J. Mentalism in linguistics. *Language*, 1964, *40*, 124–137.

Langer, S. *Philosophy in a new key*. Cambridge, Mass.: Harvard University Press, 1942.

Lerner, M. *The mind and faith of Justice Holmes*. Boston: Little, Brown, 1943.

Medawar, P. B. The science of the possible. *Listener*, 1970.

Meltzer, D. *Sexual states of mind*. Roland Harris Educational Trust Library. Perthshire: Clunie Press, 1973.

Meltzer, D., Bremmer, J., Hoxter, S., Weddell, D., & Wittenberg, I. *Explorations in autism*. Perthshire: Clunie Press, 1975.

Ouspensky, M. D. *In search of the miraculous*. New York: Harcourt Brace, 1949.

Russell, B. *An inquiry into meaning and truth*. New York: Norton, 1940.

Schilder, P. *Mind. perception and thought*. New York: Columbia University Press, 1942.

Twaddell, W. On defining the phoneme. *Linguistic monographs, 1935, 16*,

Whorf, B. L. An American Indian Model of the universe. In J. B. Carroll (Ed.), *Language, thought and reality*. Cambridge, Mass.: M.I.T. Press, 1956, pp. 57–64.

Whorf, B. L. Language, mind and reality. In J. B. Carroll (Ed.), *Language, thought and reality*. Cambridge, Massachusetts: M.I.T. Press, 1956, pp. 246–270.

Wittgenstein, L. *Philosophical investigations*. Translated by G. E. M. Anscombe. Oxford: Basil Blackwell, 1953.

19

Continuities and Discontinuities in Language Development over the First Two Years

Paula Menyuk

Division of Reading and Language Development, Boston University

There are two distinct theoretical positions concerning the development of language over the first two years of life. The first, primarily held by linguists, claims that development over this age span could be discretely divided into a prelinguistic and a linguistic period. During the first of these periods the infant randomly babbles all the possible speech sounds of language and is not aware that speech sound sequences have *meaning*. Following this period, there is an interval of silence. It is during this interval that the infant, presumably, becomes aware of the meaningfulness of the speech signal and then systematically begins to acquire the speech sound distinctions, both perceptually and productively, of his or her own language. Thus, early verbal behavior was conceived of as being composed of two distinct or discontinuous periods (Jakobson, 1968). In like fashion the earliest *linguistic* behavior, word comprehension and word production, was differentiated from word combination, comprehension, and production. The earlier behavior was categorized as asyntactic or arelational and the latter as syntactic or relational (Bloom, 1973).

In opposition to this position, many behavioral psychologists have claimed that the development of language over the first two years and beyond is a totally continuous process. Infants are first "shaped" to vocal-

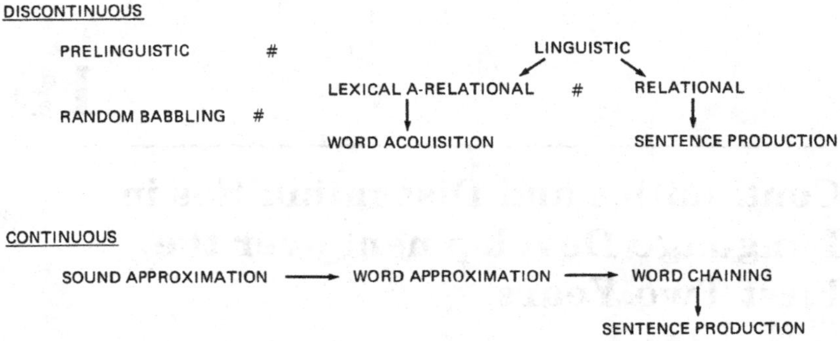

DISCONTINUOUS

    PRELINGUISTIC   #   LINGUISTIC

                                 LEXICAL A-RELATIONAL   #   RELATIONAL

    RANDOM BABBLING   #   ↓

                                        WORD ACQUISITION   SENTENCE PRODUCTION

CONTINUOUS

    SOUND APPROXIMATION ──────▶ WORD APPROXIMATION ──────▶ WORD CHAINING
                                                                        ↓
                                                             SENTENCE PRODUCTION

FIGURE 1. Theoretical positions on the issue of discontinuous or continuous development over the first two years.

ize the speech sounds of their particular language during the babbling period and then to chain these speech sounds into lexical items. Words are then chained together to form sentences. What occurs, in verbal behavior, over this age span is a closer and closer approximation to larger and larger chunks of the models provided by the environment (Staats, 1971). A graphic representation of these two opposing positions is presented in Figure 1.

There have been a number of studies of normal infants' speech sound perception and production which indicate that neither position concerning the babbling period is completely correct. Over the first year of life the infant neither randomly babbles all the possible speech sounds of human language nor begins to approximate more closely the speech sounds of the particular language. Further, the infant indicates comprehension of some of the prosodic elements of the speech signal and produces babbled utterances marked by these prosodic elements. In this way the infant communicates with others and others with the infant. Somewhat later, comprehension of words and word combinations occurs (Menyuk, 1971, chap. 3).

The frequency of the particular sounds produced by the infant changes in time so that some sounds which are produced very frequently at the beginning of the babbling period are replaced, to some extent, by other sounds later on and these by still others later. In addition, the consonant and vowel composition of babbled utterances changes in time. These systematic changes in the composition of babbled utterances have been found in studies of the babbling of infants from several differing linguistic environments. These findings indicate that the structure of babbling is, to some extent, dependent on the maturation of the vocal mechanism and changes in the posture of the infant but, also, on auditory information. The structure and composition of the babbled utterances of Down's syndrome infants (Smith & Oller, in press) is quite similar to that of normal infants, but this is not so with deaf infants (Menyuk, 1972). These latter infants continue to babble a very limited repertoire of sounds and, indeed, their vocalizations are much less frequent than those of normal and Down's infants. In addition, Down's infants mark their babbling prosodically. In another population of handicapped infants, with varying degrees of brain damage, vocalizations are quite different from birth and prosodic elements are not applied to indicate communicative intent (Ricks & Wing, 1975).

Systematic changes in speech sound discrimination behavior have also been found over the first year of life. Thus, some sounds are discriminated before others and sounds in simple contexts (CV) are discriminated before sounds in more complex contexts (CVCV). The sequence of systematic changes is apparently not totally universal. Speech sound discrimination appears to be affected by the particular cues provided by the environment at a surprisingly early age. Thus, some sounds are differentiated by all infants and others only by those infants who have been exposed to examples of these speech sound differences in the language that they hear (Streeter, 1976). On the other hand, the sequence of development of *both* perception and production of the basic prosodic elements that mark sentence intonation (statement, emphasis or demand, request) and state (friendly and unfriendly) and perception of sex of speaker have been found to be universal. The data do not support the notion of random babbling *or* the shaping of babbling by the language of the environment, and even though there appears to be an effect of particular language environment on speech sound discrimination, this cannot come about by shaping since these differences are observed in the very abstract context of CV syllables and as early as two months of life.

The data also do not support the notion of a discontinuity in development between one-word and two-word combination productions. During the end of the first year of life or the beginning of the second, recognizable morphemes begin to be produced by normal infants. Some children produce single-word utterances alone, others a single word in combination with a jargon phrase and others a combination of the two. These utterances are marked prosodically to state, demand, or request as babbled utterances were previously. The sound composition of these early words is very similar to the sound composition of early babbled utterances. The sequence of realization of speech sound distinctions in words is very similar to the sequence of usage of sounds during the babbling period. There are some data which indicate that the sequence of speech sound discrimination, when the task is differentiation of nonsense syllables, is very similar to the speech sound discrimination in early infancy and the sequence of speech sound distinctions produced in words. However, when children are asked to discriminate between real words, their performance is very much affected by familiarity with the words. What the children appear to be doing initially is attempting to discriminate between words on the basis of their meaning rather than their speech sound features. These findings imply a discontinuity in speech perception behavior between the prelexical and lexical periods (Menyuk & Menn, 1979). Such a discontinuity does not appear to exist between production of single words and word combinations.

Before holophrastic utterances are produced, observational data indicate that a small set of words is understood. Some recent studies have examined infants' comprehension of action–object relations (kiss teddy, tickle car) and comprehension of three-part relations (boy kiss teddy) during the holophrastic and sequential single word (daddy ... car) periods. It was found that some two-part relations are understood when only single words were being produced, and many two-part relations and some three-part relations understood when sequential single words are being produced. Thus, some words and phrases are understood in *context* before recognizable morphemes are produced, some two-part relations out of context during the holophrastic period and some three-part relations when the child is still not producing smooth word combinations (Menyuk, 1977, chap. 2). These data do not imply a discontinuity in language processing between one-word and two-word utterance production.

The similarties in sequence of development of language behavior in normal and Down's syndrome children disappear at the point at which the normal child becomes aware of the meaningfulness of the speech sequence. This step in language development will occur for the Down's syndrome child at a much later age. As indicated previously, the similarities in the language behavior of deaf and hearing infants cease to exist at a much earlier period, probably at about 6 months of age when true babbling begins to occur. The sequence of development of production and perception discussed above is outlined in Table 1.

These findings with normally developing children and children with developmental anomalies indicate that there are both continuities and discontinuities in language development and that they are the product of the varying factors that bring about particular language behaviors. During the prelexical period speech sound discrimination and production develop in a certain order. In addition, there is comprehension of aspects of pros-

TABLE 1. Developmental Sequence in Perception and Production

Perception	Production
Difference of syllables	Vocalization
Affective prosody	Cooing
Difference in multisyllables	Babbling contrasts plus prosodic contrasts
Difference in sentence contour	
Comprehension of words in context	Syllable length utterances
Comprehension of some two-part relations	Words marked prosodically
Comprehension of many two-part relations	Sequential single words
Comprehension of some three-part relations	Word combinations

ody and production of some of these aspects in a certain order. These aspects convey communicative intent and affect. What appears to be necessary for the development of these behaviors are intact vocal and auditory mechanisms, the ability to organize and relate the products of these mechanisms, and the basic ability to interact with others. Thus, normal and Down's syndrome infants display very similar patterns of language development during this period, whereas deaf infants and some brain-damaged infants do not, but for differing reasons; the former because they cannot hear themselves or others and the latter because they are unable to organize auditory input and vocal output, relate the two, and engage in interaction with others.

Some aspects of these early behaviors are carried over into the lexical period and are, therefore, continuous. Speech sound realizations are mapped into words. Prosodic elements continue to be used with lexical items in the same way they were used with babbled utterances and for the same purpose; that is, to convey communicative intent. However, the ability to become aware of the meaningfulness of utterances in context requires other competences in addition to intact auditory and vocal mechanisms and the ability to relate to others. This conceptual competence is not available to the Down's syndrome child at the same age as to the normally developing child. A much longer time is needed to develop this competence. With other children, this competence is never achieved. On the other hand, the deaf child who is exposed to language in a mode other than auditory (i.e., sign language) displays the same sequence of language development in this other mode as is observed with the normally hearing child.

Once having comprehended the meaningfulness of utterances and the relations they express, development once again becomes continuous. The normal child comprehends two-part and then three-part relations. Single

	PRELEXICAL	LEXICAL AND RELATIONAL
SPEECH SOUND REALIZATION		→
PROSODY COMPREHENSION AND USE		→
SPEECH SOUND DISCRIMINATION	SYLLABIC —#——→ MORPHEMIC	
WORD COMPREHENSION	#	→
WORD PRODUCTION	#	→
RELATION COMPREHENSION	#	→
RELATION PRODUCTION		# →

FIGURE 2. Continuities and discontinuities in aspects of language development over the first two years.

words, sequential single words, and then two words are produced. There does not appear to be a discontinuity in language development between one-word and two-word production. However, later acquisition of the specific structural rules of the language may require still another type of competence. Figure 2 presents the proposed continuities and discontinuities between the prelexical and lexical periods discussed above.

Language development, like other aspects of development, is composed of continuities and discontinuities. Determiniation of the *causes* for such discontinuities and continuities with both normally developing and handicapped children is the most important aspect of the issue.

REFERENCES

Bloom, L. *One word at a time.* The Hague: Mouton, 1973.
Jakobson, R. *Child language, aphasia and phonological universals.* The Hague: Mouton, 1968.
Menyuk, P. *The acquisition and development of language.* Englewood Cliffs, N. J.: Prentice-Hall, 1971, Chapter 3.
Menyuk, P. *Speech development.* Indianapolis: Bobbs-Merrill, 1972.
Menyuk, P. *Language and maturation.* Cambridge, Mass.: M.I.T. Press, 1977, Chapter 3.
Menyuk, P. & Menn, L. Early strategies for perception and production of words and sounds. In P. Fletcher & M. Garman (Eds.), *Studies in language acquisition.* Cambridge, England: Cambridge University Press, 1979.
Ricks, D. & Wing, L. Language communication and the use of symbols in normal and autistic children. *Journal of Autism and Childhood Schizophrenia,* 1975, 5, 191–221.

Smith, B. & Oller, D. K. A comparative study of the pre-meaningful vocalizations produced by normal and Down's syndrome infants. *Journal of Speech and Hearing Disorders,* in press.

Staats, L. Linguistic-metalinguistic theory versus an explanatory S–R learning theory of language development. In D. Slobin (Ed.), *The ontogenesis of grammar.* New York: Academic Press, 1971.

Streeter, L. Language perception of two-month-old infants shows effects of both innate mechanisms and experience. *Nature,* 1976, *259,* 39–41.

20

On the Counterverbality of 'Nonverbal' as a Verbal Term

John B. Newman

Professor Emeritus, Department of Communication Arts and Sciences,
Queens College of the City University of New York

Those who have studied the folklore of language must certainly have encountered 'popular etymology',[1] in which "an irregular or semantically obscure [linguistic] form is replaced by a new form of more normal structure and some semantic content—though the latter is often far-fetched" (Bloomfield, 1933, p. 423). So, for instance, "forlorn hope" came into English because the Dutch expression *verloren hoop* (which means "lost troop") looked (to the eyes of readers of English) as though it meant "forlorn hope". Popular etymology may also be illustrated in another way. We could point out (and, we daresay, without untoward redundancy or without restating what might be presumed to be "common knowledge") that 'noisome' is not a variant of 'noisy', that 'fulsome' (despite its looking like 'handsome', 'toothsome', and 'winsome') is not a synonym for 'lavish', and that 'machination' has nothing to do with machines. These words may LOOK or SOUND (or be made to sound) like well-known, similar-looking words. Common sense (whose commonality, despite the order of its words, commonly comes AFTER the sense) then "tells" us (i.e., presumably forewarns us) that such "strong family resemblances" in the appearance of these words must indicate equally strong family resemblances in their meaning as well. But there really is no justification (beyond popular belief—and, hence, popular etymology) to assume that because of simi-

larities in their appearance (i.e., their surface structure) words must somehow be etymologically or semantically related in their deep structure as
well. Mark Twain summed all this up in his etymology of 'Massachusetts'.
It was, he insisted, derived from 'Moses'. Over the years, he explained, the
"oses" somehow got lost, and there was added the "assachusetts"!

Language being what it is, however, nothing ever stays that simple.
Which reminds us of something that Paul Anderson once said: "I have yet
to see any problem, however complicated, which when looked at the right
way did not become still more complicated."[2] Consider, for instance, the
word 'verbal.' It means "of, pertaining to, or associated with words," but,
as *The American Heritage Dictionary of the English Language* (Morris,
1969) goes on to explain, " 'verbal' (adjective) is less precise than 'oral' in
expressing the sense of "by word of mouth." 'Verbal' can also refer to what
is written; 'oral' cannot." The dictionary (that is, the 1969 edition of *The
American Heritage Dictionary*) cautions that 'verbal' can refer to what
is written as well as to what is 'spoken,'[3] but users of the term do not seem
to have always heeded the warning. 'Verbal' has blended into and become
identified with 'vocal.'

No lesser personages than P. F. Strawson, the philosopher, and
Edmund Leach, the anthropologist, have exhibited this diversity in the
use of the term in their writings IN THE VERY SAME VOLUME (Minnis, 1971).
In an essay entitled "Meaning, Truth, and Communication," Strawson
writes, "An utterance . . . need not be vocal; it could be a gesture or a
drawing or the moving or disposing of objects in a certain way" (p. 92).
This description would conform with the dictionary definition quoted
above. But later in the very same volume, in an essay entitled "Language
and Anthropology," Edmund Leach explains that "in a purely mechanical
sense, a verbal utterance is simply a pattern of sound superimposed upon
breath" (p. 140).[4]

Words are, it is true, most often communicated vocally. So it seems,
then, that it has simply come to be assumed that a person who communicates verbally—that is, a person who communicates by means of
words—does so vocally. This would, then, lead one to believe that a person may be said to be "verbal" when he or she[5] is speaking and
"nonverbal"[6] when not speaking.[7] The consequence of this curious lack of
semantic precision in the case of the word 'verbal' has had an even more
curious effect on the meaning and usage of its converse: the word
'nonverbal'.

To begin, a word is usually thought to be a unit—or, at least, an
item—having a specific meaning in the lexicon of a language, regardless
of the derivation of that meaning. Thus, a hat is worn on the head, a shoe
on the foot. One sits on a chair and rests one's elbows on a table; one does

not sit on a table (although one can if one is obstinate) and rest one's elbows on a chair (reason WILL prevail!—unless one insists on displaying one's unusual physical agility). But there are some words that are popularly used in a great variety of situations with little or no regard for their specific meanings. One could speculate that these words must have achieved their currency (and their vacuity) precisely because they actually had no specific meanings, and merely served as fillers when the speaker or writer was fishing for a word and just could not seem to hook one. But there is no need to speculate about this. The fact is that lexical items of this sort simply run counter to the function presumed of ordinary words. Rather than darting like little arrows of meaning shot from the speaker's lips into the ears of hir listener(s)—(as it were!)—these vacuous lexical fillers (!) billow forth like a verbal smokescreen, masking such morsels of meaning as may have somehow managed to make it through the murk. Nor is there anything nefarious about all this. The fault, for a change, is not the speaker's. It's hir language!

Oops! . . . Hir WHAT?

Hir vocabulary, hir rhetoric—you know—hir communication. It's— you know—a fault in the structure of the code. And so, if the speaker is not careful in phrasing hir thoughts, hir meaning can be misconstrued.

Now, dear reader, if you think you have heard this excuse before (if the previous paragraph sounds vaguely familiar), that is because this rationalization, this excuse for what passes as an "explanation," does not consist of "real" words[8]—i.e., lexical items that have specific meanings— but, instead, is made up of counterwords—i.e., references that are loosely used in a great variety of situations and with ever vaguer meanings. Counterwords comprehend so many things that they can mean virtually anything to virtually anyone about virtually anything. Not only do such everyday glyphs as 'nice' and 'awful' (as the dictionaries tell us) exemplify counterwords, but even so objective (if not actually technical) a term as 'semantics' can undergo a similar lexicological metamorphosis.[9] The same is true of the word 'nonverbal.'[10] And so, since it is a counterword, 'nonverbal' (which is a term which is itself verbal) has counterverbality.

Adam Kendon (1978, p. 90) goes even further, and says, straight out, that "'non-verbal communication' is a non-topic."[11] The author of the book (Key, 1977) Kendon was reviewing when he made that statement, like so many others who have written and are writing on the subject (even including us, here, now!), spells 'nonverbal' as an unhyphenated, positive orthographic entity, which makes the term appear to have provenance as a lexical item, and, hence, have an apparently legitimate claim toward being a "real" word (that is, one having a specific meaning, like 'hat' and 'shoe' and 'table' and 'chair'). But regardless of the presence or the absence of the hyphen, the term 'nonverbal communication,' as Kendon would have it, is extremely unsatisfactory as a name for a phenomenon or a field of study:

> It has always seemed quite wrong to try to specify a field of study by stating what it is not. Suppose I were to draw up a list of titles on "non-avian zoology" or on "non-Nineteenth Century literature." This would obviously be absurd. There would be no possible coherence in such an attempt.... [And] a comprehensive bibliography on "non-verbal communication" is little better. (p. 90)

"To add inconsistency to absurdity," Kendon goes on to say:

> It is impossible to deal sensibly with words on the one hand, and every-thing else that plays a part in communication on the other. As Ray Bird-whistell once said, to talk of "non-verbal communication" is like talking about "non-heart physiology." Communication in face-to-face encounters, at least, is a complex phenomenon in which all aspects of action are involved. Any attempt to analyze what is going on apart from words is doomed from the outset.

Yet in Mary Ritchie Key's bibliography on "nonverbal communica-tion," Kendon tells us, there are listed:

> Studies on rates of eye blinking and extra-terrestrial communication; studies on the specialization of function in the brain's hemispheres and studies of gestures; studies of the physiology of breathing and studies of baby talk; studies on glossolalia, on the drum languages of Africa, on dance, [and] on pantomime.... [There are] many references on phonetics and a good many on various aspects of sociolinguistics and conversational [organization and] analysis.... There are references to studies of snoring, and similarly curious yet neglected actions, which often have important effects on others who witness them and so should certainly be grist for the mill of any student of communication.... In other words, what we are given is a listing of titles that are relevant, in the broadest possible sense of "rel-evant," to understanding how people manage to communicate with one another when co-present. (pp. 90, 92)

Any term whose domain of meaning or field of reference includes (or even implies) half as much as all this is unquestionably a counterword. For if so much can pertain, just what is meant if something is said to be "non-verbal"? Yet, despite it all, Kendon points out that there are not included in Key's bibliography on nonverbal communication any references to "anything that is concerned with communication by way of the written word."

The omission was not an oversight. The word 'nonverbal' simply means one thing to Key, but quite another thing to Kendon. Simply because Kendon and Key both arrange a string of letters in the same way and pronounce the result identically is no reason to assume that they are both using the same "word" to comprehend the same semantic domain. And it is our contention in this essay that they are not.

The letters 'n-o-n-v-e-r-b-a-l,' when strung together as a "word," constitute a homonym, which (according to *The American Heritage Dictionary*—again!) is "one of two or more words that have the same sound and often the same spelling but differ in meaning." We hold that 'n-o-n-v-e-r-b-a-l' represents two separate and different words, both of which have the same spelling and both of which have identical pronunciations; but they differ in meaning.

To Key, 'nonverbal' means "that which is not spoken"; to Kendon, 'nonverbal' means "that which is not said." The difference between the two is that 'spoken' is restricted to vocal utterance, while 'said' means "verbalized" or "put into words, regardless of whether the words are spoken, writen or otherwise made manifest." Kendon and Key seem to agree that that which is "nonverbal" is *not verbal,* but they do not seem to agree as to what IS "verbal." Could there be a more classic instance of counterverbality?

Consider, now, what Kendon was referring to in his statement above concerning "non-avian zoology." A zoology of the hairy quadrupeds[12] would obviously be nonavian, but it should not therefore be considered for the birds—or against them(!).[13] A nonavian zoology would simply be another zoology, one that is different from a zoology of the birds. By the same token, that which is "nonverbal" is neither "proverbal" nor "antiverbal"; it is simply different from "verbal" (whatever "verbal" may eventually be decided to mean).[14] To consider "nonverbal" as the negative of "verbal" is to be party to popular etymology. 'Nonverbal' LOOKS like the negative of 'verbal' (and, at one time, it may well have been). But meanings change; and many writers no longer consider 'nonverbal' as the negative of 'verbal,' but rather as signifying aspects of communication that are separate and different in some regard from those of verbal communication.[15] Key apparently thinks so, and Kendon (who still holds to the negativity of 'nonverbal') is unfair (at least!) in taking her to task therefore.

When Ludwig Wittgenstein (1976) declared in the *Tractatus Logico-Philosophicus* that "that which can be said, can be said clearly; and whereof one cannot speak, thereof must one be silent,"[16] he set up a dichotomy that equated effability with speech and ineffability with silence. This epitomizes the verbal-nonverbal dilemma we are trying to resolve here. But Wittgenstein was innocent of any wrong-doing—in this instance, at any rate. The fault lies in language itself.[17] Words are used figuratively as often, if not more often, than they are used literally. (This, indeed, is the basis for their inherent counterverbality.) When Wittgenstein said "whereof we cannot speak," he was referring to far more than any ephemeral, temporary, or chronic disability to pronounce the words that one may try to put together into sentences. Wittgenstein's phrase refers to that which it is not possible to put into words, whether written OR spoken, whether sounded or silent—indeed, that which may not—in fact, cannot—even be thought of.[18] Furthermore, the "silent" in "thereof must one be silent" is not restricted to the lack of sound or the failure to speak. Someone who has been silent may be someone who has not been "heard from." Writers who have not produced, for instance, may be said to have been "silent," as well as, of course, speakers, Signers (that is, Sign Language users), fingerspellers, and even mimers. The confusion here stems from the counterverbality of the word 'silent.'

Alfred Korzybski, the founder of General Semantics, went even further in this regard. He blamed those who were "semantically benighted" for constructing:

> a delusional world in which the subjective products of the nervous system are systematically identified with [the objective level of] the scientific reality upon which survival depends. This way lies madness—the peculiar madness of the highly verbalized civilization to which we belong. The way back to mental health is . . . achieved in . . . the conviction that reality is nonverbal. (Black, 1949, p. 239)

The objective level of the scientific reality upon which our survival depends, Korzybski (1941) tells us,

> is not words, and cannot be reached by words alone. We must point our fingers and be silent, or we shall never reach this level. Our personal feelings, also, are not words, and belong to the objective level. (p. 399)

If, then, 'nonverbal' comprehends not only all that is involved in how people manage to communicate with one another when copresent, but their personal feelings as well, then the term, by referring to so much, actually serves for very little, if any purpose, in helping us understand what it purports to mean. We shall argue, therefore, that the verbal term 'nonverbal' be set aside because of its counterverbality, and that an entire set of new terms be adopted to refer specifically to each of the several aspects that may pertain in a particular communicative situation.

Linguists generally have rejected Alfred Korzybski's theories of language as he set them forth in the system he chose to call "General Semantics."[19] The inadequate solutions that his theories afforded and their excessive logophobia are generally given as the reasons for their rejection. "Nevertheless, he did isolate a logical contradiction: language is supposed to communicate experience, yet by its very nature it is incapable of doing so." (Farb, 1974, p. 170) And, indeed, as it is usually formulated, verbal language imperfectly fulfills the function it is supposed to fulfill: which is to approximate the thoughts, feelings, awareness, or needs of the speaker or writer. Anyone who intends to express IN WORDS any of these aspects of hirself (or anything related to, or dependent upon, any of these aspects) must make up for the deficiency in some way OTHER THAN by simply using words.

It is no longer fashionable among professionals (and other experts) to refer to the process of speech as being an "overlaid" or "secondary" biological function (Gray & Wise, 1959, pp. 137n, 474–476).[20] But the fact remains that human beings do make use of physiological mechanisms whose primary vegetative biological purpose is to take part in the processes of respiration, mastication, and/or deglutition in order to produce those "noises made with the face" (Smith & Ferguson, 1951, p. 1) that we call speech. The so-called "vocal organs" (i.e., the larynx, tongue, teeth, lips, and lungs) are also all to be found in many animals that do not have speech; all these animals use these organs for the more "basic" biological purposes of respiration, mastication, and deglutition. Hence, human beings, who do have speech, are said to have added or "overlaid" the function of speech on these organs.

But such a view overlooks the high degree of evolutionary development in human beings. Not only have the human larynx, tongue, teeth, lips, and lungs achieved high degrees of specialization in themselves, but, more important, so has the human brain. In fact, the human brain has achieved such incredible complexity as not only to permit a fine control and minute synchronization of the various muscles brought into play in speaking, but also to permit the retention of the rules of grammar and the vocabulary of language as well. Philip Lieberman (1975) goes so far as to say that some of the early hominids appear to have developed a communication system that retained a mixed phonetic level that relied on both gestural and vocal components, while others (including, presumably, the forerunners of our own species) "appear to have followed an evolutionary path that has resulted in an almost complete dependence on the vocal component for language, relegating the gestural component to a secondary 'paralinguistic' function." One of the consequences of this development, Lieberman goes on to say, is that the adult human larynx is not as efficient for respiration as, say, the larynx of the horse, nor is it as efficient as that of the Australian mud fish in sealing off the lungs (and thus preventing asphyxiation from choking on improperly swallowed food). But "the development of the 'bent two-tube' supralaryngeal vocal tract," in which the larynx of the adult human exits into the pharynx (rather than directly into the oral cavity, as in the case of nonhuman primates), has made it better adapted for efficient phonation rather than for any "primary" vegetative biological function deemed to be more "basic" than the production of speech.

Still, we must go on breathing (and, at times, chewing and swallowing) even while producing those noises we make with our faces. And so there are always muscle movements, many of them visible—if not overtly obvious—that are going on while vocalized speech is taking place. Thus, in a sense, we are always gesturing (at least with our faces) while we are speaking. It should not, therefore, be at all surprising that these head and face movements, perhaps arm and shoulder movements as well, should have become incorporated into the act of speaking itself. We have undoubtedly heard it said of voluble individuals, particularly those of Mediterranean and/or Latin heritage, that "if their hands were tied, they would be struck dumb—unable to say a word." Well, such a characterization is nothing more than an ethnic slur. The fact of the matter is that anyone, regardless of ethnic background, whose hands were tied would find hir speech would indeed be impeded. We speak as much with our arms, heads, shoulders, faces, and hands as we do with our lips, tongues, and vela.[21] And these gross, visible movements of our bodies not only formulate and transmit messages on their own, in and of themselves, but they also serve to emphasize, clarify, and otherwise compensate for whatever it may be felt is lacking in the words that make up the sentences that constitute the verbal language produced in the act of speaking.[22]

Before going any further, however, we should like at this point to respond to a comment made by Professor Birdwhistell, the creator of kinesics, the study of body motion communication, in which he decries "an atomistic and mentalistic model of a human being as a thing in itself" because "with such a model we are condemned to do our research on little balloons full of words which are somehow framed or filled out by gesticulations which we could dignify although not clarify by calling them 'nonverbal communications'" (Birdwhistell, 1970). Let us make it perfectly clear that we have no intention of arguing with Professor Birdwhistell in this (or in any) regard. It is reassuring to know that he, too, agrees that calling some things "nonverbal communications" may dignify them without clarifying them. But we feel that in introducing a complex and involved subject, it is helpful to lead into the discussion gradually—from the well known and the familiar to that which (because it takes place largely out of our awareness) seems to be arcane and esoteric. No one can talk about everything at once, and, like trousers, which are best donned one leg at a time, complex ideas are best presented one premise at a time. We understand what Professor Birdwhistell is saying here, and we do not disagree with him. We do, however, promise to try not to reify (much less deify) this, or any other, explanatory paradigm.

In *The Unspoken Dialogue: An Introduction to Nonverbal Communication,* Judee K. Burgoon and Thomas Saine (1978) say that "authorities have identified at least six ways in which nonverbal communication works in conjunction with verbal messages." These ways are listed, and explicated, much as follows:

1. *Redundancy.* Nonverbal cues can be used to say the same thing that is being said verbally. When we tell people that we are pleased with them, we smile to repeat our message. At a noisy hamburger stand, we may hold up two fingers while we yell our order. This built-in repetition increases the accuracy of the communication interchange.

2. *Substitution.* Sometimes, rather than say something aloud, we signal it nonverbally. Rather than tell a friend we are sorry about her misfortune, we may pat her on the shoulder. At the movie theater, we may simply hold up four fingers when asked how many tickets we want. When we are angry, we may simply walk away instead of trying to verbalize our feelings. Because of the richness of nonverbal imagery, it is often more effective to substitute a nonverbal cue for its verbal counterpart: obscene gestures have their impact in part because of the visual images they create. The substitution of nonverbal cues can also increase the efficiency of communication. A nonverbal message frequently can be conveyed much more rapidly than a verbal one, and two or more messages can be delivered at one time.

3. *Complementation.* We often use nonverbal cues to supplement or modify what is being transmitted verbally. The nonverbal elements simply add more details, but in this case the details are neither identical to, nor a replacement for, a verbal message. They serve to expand on the message being conveyed. While reporting an accident he or she has witnessed, the individual animatedly recreates the scene through gesture, posture, and voice. All these elements make the story more vivid. They also reveal how that individual feels about the accident. Similarly, when we are asked to reprimand someone or pass along bad news, our nonverbal cues may reveal that we are reluctant to undertake the task. The nonverbal complement is generally important in clarifying both the verbal message itself and the nature of the relationship between speaker and receiver.

4. *Emphasis.* Often we use nonverbal cues to accentuate or punctuate what we are saying verbally. Vocalic elements are especially useful for this purpose. We may moderate our pitch or rate of speaking to emphasize a point. Gestures are also commonly used for this purpose. . . . Much humor is dependent for its success on the use of nonverbal elements. Timing, posture, and facial expressions all may be used to emphasize the punch line.

5. *Contradiction.* We may frequently use nonverbal cues to send messages that conflict with the verbal message. Sarcasm depends on just this contrast. The vocal inflection signals that the verbal message means the opposite of what it says. . . . On a job interview, we may signal with our body that we are very interested while our words are more cautious. When we are

angry or hurt by someone, we may say, 'Nothing is the matter', while our voice and body cry out that something is bothering us. How many times have people admired something you created or claimed to agree with you, when you were absolutely certain that their opinion was the opposite? Chances are, they did not do a very good job of masking their true feelings; their nonverbal behaviors gave them away.

6. *Regulation.* Nonverbal elements regulate communication interactions. Through our nonverbal behaviors, we can govern when a person talks, how long she or he holds the floor, and even the topics discussed. Or we can use nonverbal signals to gain control of the communication ourselves. We can prevent others from interrupting us and can assure their attention. (Burgoon & Saine, 1978, pp. 10–13)

According to this scheme, then, the same particular signal (or "nonverbal message") can serve either to reiterate, or to substitute for, or to complement, or to emphasize, or to contradict another signal (the "verbal message"), as well as to regulate the interpersonally directed behavior of a communicative correspondent. Yet, according to this scheme, that same signal is still referred to (how polysemic can we get?) simply as a "nonverbal message." Talk about counterverbality! Would it, then, not be at least somewhat more sensible to refer specifically to each function individually by its own designation? Thus, an action (or a circumstance) used to reinforce a verbal statement by repetition will be differentiated from a similar (if not the same) action (or circumstance) used as a substitute for a verbal statement, and so on. When, and if, such actions (or circumstances) happen, of themselves, to be the same, or of similar nature or structure, that would be all the more reason to distinguish them, one from the other. And even if they happen to be different—in nature, in kind, or in structure—simply to lump them all together as "nonverbal communication" is not only to obscure and to obfuscate the problem, but also to fail to recognize essential differences, purposes, and intents in the process.

Since it is essentially a matter of differentiating what is usually designated as the verbal from the nonverbal, we propose to retain the "verbal" basis in developing a nomenclature for the taxonomy we wish to propose here. Thus, *verbal* will remain the designation in adjectival form of that which is based upon, or has to do with, words and/or word-elements, and *verbal communication* shall remain the designation of any and all sorts of interaction consisting of, or based upon, words and/or word-elements, regardless of the mode, the channel, or the manner of such formulation, expression, transmission, reception, interpretation, or understanding. Any interaction, or any aspect or any part of any interaction, in which words and/or word-elements are NOT used (heretofore generally referred to simply as "nonverbal") will hereafter be specifically designated as to purpose, intent, and fulfillment. The basis for differentiation, however, will remain pegged to the *verbal* and the concept of *verbality*. One reason for insisting upon this distinction right at the outset is to avoid unnecessary blending or fading into *language* and that which is deemed to be *linguistic*. Without going into it any further at this point, let us account for the distinction merely and simply by indicating that we deem *language* and that which is *linguistic* to consist of far more than mere *verbality* and that which is merely *verbal*. With this in mind, let us, then, proceed.

Somewhere (we have forgotten exactly just where) Birdwhistell once wrote that he was able to recognize and isolate over one hundred separate channels of nonverbal communication. Like the lines of light on the screen of a television set, says Birdwhistell, human beings communicate with each other in all these channels, all of which together form an image, much like the lines of light on a television screen do, an image which "constitutes" the "person" doing the communicating. That is certainly a vivid conception of the process! We have no intention of reconstructing or even referring to all these many "lines" here, but would like to offer a sample— in fact, a dozen. These 12 are a combination of the reports of Egolf and Chester (1973), Burgoon and Saine (1978), and our own version. These 12 forms that nonverbal messages take fall into three general groups, as follows:

The first, which we shall classify as "body actions," consists of:

1. *Kinesics*. This comprehends all motions of the body, including postures, stances, attitudes, gestures, expressions, and what-all-ever that are in any way involved in (or that take place as part of) the process of communication. Such a description involves such a vast diversity of actions and motions by the body as a whole, as well as by the various members of it, that it should not be at all surprising that the area of kinesic studies has been broken up into several subordinate areas.

2. *Oculesics*. This refers specifically to the use of the eyes in communication. That would comprehend all movements of the eyes, including winks and blinks, all sorts of gaze contacts and fixations, as well as gaze avoidances.

3. *Haptics*. This refers to all sorts of touching behaviors, which are, by their very nature, communicative. These would include handshaking, backslapping, pinches, jabs, pokes, pats, strokes, hugs, and kisses of various sorts and kinds. Haptics would include within its province of concern all indications of intentions to touch in these various fashions, even though the actuality remains unconsummated. Thus, haptics would include such actions as are meant to indicate a touch of some sort, like "throwing" a kiss, feinting a jab in the ribs, or making out as though to pinch a cheek.

4. *Olfactics*. This refers to personal odors, both internal and external, both natural and applied. This is a whole subject unto itself to which we can only allude to here. Anyone who may be interested in pursuing this subject further might well begin with Ruth Winter's (1976) *The Smell Book*.

5. *Vocalics*. This refers not only to the effects of voice quality on communication, but also that of prosody, including rate and rhythm of articulation and pronunciation, the clarity of enunciation, inflectional and intonational patterning, and the length and frequency of silences, hesitations, and "drags" of vocalization in the flow of speaking.

Another group of the forms that nonverbal messages can take, we shall classify as "personal appearance." That would include:

6. *Organismics*. This refers to the effect on communication of the body size, structure, and shape of the participants in the process. Whole volumes have been written on this subject, and it would be sufficient for our present purposes to simply remind (if not to refer) the curious reader to the works of Sheldon (1940, 1942) on body-types and temperament. What is closer to our present purposes, however, are the more immediately notable facts of body size. To encounter a person of great height or of great bulk, or, conversely, to interact with a person who is very short, very thin, or very tiny is undoubtedly to have an effect on one's communication with such a person. Who has not felt uncomfortable, inadequate, perhaps even guilty, in the presence of a physically handicapped individual, whether or not he or she is in a wheelchair? This is the area of concern of organismics. But there are other aspects of this that pertain to communication and that have nothing to do with the extraordinariness of the person's physical appearance. These are age, sex and gender, and skin color.

All societies are aware of age groupings, and human relations are fundamentally affected by the cultural constraints imposed upon the relations that may, should, or ought to take place between members of different chronological groups. There can be no question but that these constraints affect communication.

Human beings are sexually dimorphous: structurally, they are either male or female.[23] But, depending upon a great number of factors, human beings are of many genders. Anatomically, a person may be shaped like a male, but socially, personally, or only in certain regards, habits, or mannerisms may act or otherwise disport hirself like a passive or an aggressive female, etc., etc. All of these factors unquestionably affect the individual's communication with others.

Americans, from virtually the time they were first called Americans, have felt guilty about their feelings about skin color as a factor in human relations. But the problem seems to have become (if it was not always) worldwide. We talk to people's skins at least as much as we do to their age (or our perception of their age), as much as we talk to their sex (if not their gender), as much as we do to their organismic size and shape. People with darker skins (the varieties and intensities of the hues are unquestionably variables in the formula!), regardless of their facial features, are addressed differently (whether the speaker is aware of that difference or not). A person may look like a Caucasian physiognomically, but if hir skin is the color of a black's, most (if not all) people will react differently from the way they would to a person who looks like a black physiognomically but has the skin color of a Caucasian. Nor does it stop there.

Consider, then, the combinations and permutations of age, sex, skin color, and body size, shape, and structure to get an inkling of the background effect of organismics on communication.

7. *Cosmetics.* Though the term usually connotes face-paint of one sort or other, cosmetics here refers to every sort of body decoration, addition, modification, and adornment affected by human beings. This would, of course, include every sort of clothing, hair styling, and prosthetic device, like canes, umbrellas, purses, and eyeglasses. When thought of in this way, cosmetics—which is not necessarily only intended to make one appear pretty, attractive, or even "presentable"—comprehends a great deal of THE PERSON. Not only does cosmetics thus include all sorts of facial and body painting and tattooing, but all sorts of cranial and facial hair styling and coloring; every kind of clothing, regardless of whether it is intended to be merely protective (like shoes to protect the soles of the feet from rocks and stones, hats to protect the head and face from sun and rain, or furs to protect parts of, if not the entire, body from the cold), indicative of social station (like a uniform), or decorative (like a costume). Further, such additions to the body need not necessarily be visible to be considered cosmetic, for corsets and girdles of various sorts, including brassières and jockstraps, as well as all other supporting undergarments are also included. And all this does not even mention such adornments as neckrings, neck chains, and necklaces; bracelets, armbands, anklets, and hip chains; earrings, fingerrings, noserings, and toerings; medals, medallions, and amulets of all sorts; and what-all-else that people hang on, or attach to, themselves.[24] Nor does it account for such props as watches, purses, pipes, cigarette-holders, and the other accoutrements and accessories that people carry around from day to day, every day, almost as a part of their corporeal selves.

A third group of the forms that nonverbal messages can take, we shall classify as "circumstances" (for, indeed, these phenomena constitute what are commonly referred to as "circumstances"!). These include:

8. *Proxemics.* Edward T. Hall (1966, p. 1) coined this term to refer to the study of "social and personal space and man's perception of it". 'Proxemics' has proliferated in meaning so that it now also refers to the distances at which people habitually stand (or sit) apart when in each other's company, and to the angles at which people interact with each other when communicating with each other. 'Proxemics' also refers to the entire question of territoriality, or the appropriation and protection of space deemed to be, or granted as being, private, or one's personal possession— as is evident in such oft-heard statements as "that's my seat!", "I'm not allowed to go in there", "that's the ladies' room" (or "the boss's private office"), etc., etc. Seating arrangements, though only a relatively minor manifestation of territoriality, have, on occasion, affected the welfare of virtually the entire world, as in the case of the hassles over seating arrangements at international peace conferences. Personal space and interpersonal distance, whether actual or perceived, communicates as much as, if not more than, words in many instances.

9. *Chronemics.* Hall (1959, pp. 23–42) wrote about "the voices of time" long before he coined the term 'proxemics.' Who arrives early (and how early) and who arrives late (and how late) to an appointment "says" a great deal about the social status of the participants and what each thinks of the other and their meeting at that time in that place. Like proxemics, 'chronemics' has also proliferated in meaning so that it not only includes time, but also the number of times as well as timing. Time, of course, refers to the moment of occurrence: the time of day, week, month, or year. Something said in the morning may not be appropriate in the evening, and what pertains in the winter will not have much meaning in the summer, and so on. But 'chronemics' also refers to the number of times a statement or a behavior or an event occurs and recurs (and recurs . . .), and so on. Is something said over and over and over again, or is the utterance a singular event whose occurrence is not only noteworthy but even memorable? Time, then, is to be distinguished from (number of) times, though both fall within the purview of chronemics. Timing, on the other hand, is a matter of effect. To say something at the exactly correct moment, after an increasingly tense silence perhaps, is the secret of all theatrical performance, particularly of the delivery of a punch line (whose punch is only as powerful as the effectiveness of the timing). So timing, which is to be distinguished from the record of occurrence (or time), as well as the frequency of occurrence (or the number of times), is also an aspect of chronemics.

10. *Artifactics.* This refers to the appearance, the furnishing, or the layout of the place in which an event occurs. Artifactics includes not only the background and the ambiance of the place, but its lighting as well.

Obviously, that which is said in the intimacy of a dimly lighted booth in the back of an intimate cafe (with violins in the background, of course!) will have a different effect (if not some sort of distinguishable "meaning") from that which is shouted while crossing a busy city intersection in the glare of high noon. But artifactics is not limited to the larger aspects of scenery and lighting, but pertains to what may be referred to as interior decoration as well: the style of the furniture (if there is any) and its arrangement, the "art" work in the place (if there is any), indeed, the very structure of the building or the natural "architecture" of the place in which the event is taking place can also be considered as artifactic. Furthermore, there are to be considered in this regard what may be thought of as the trinkets, gadgets, and doodads with which people surround themselves. The differentiation of these items from what have already been referred to as "props" may be imbricated, but, then, none of these groupings is discrete or separate from the others. Thus, a pair of spectacles may have as their manifest purpose the aid or improvement of vision, yet they may be used at least as frequently (if not more so) as a pointer, a necklace, a tiara, a swagger stick, or even a scepter. Spectacles, rings of keys, pens and pencils, umbrellas, newspapers, and purses are all items that are evident in communication by means of artifacts.

Egolf and Chester (1973) list "objectics" as a category of nonverbal communication, describing it as referring to "the effect of objects upon communication and the creation of objects by communication." We think that this category should be divided into two because of the separate impact of each on human relations. Hence, there would be:

11. *Anthropomorphization.* This would refer to the treatment of objects or infrahuman beings as though they were persons. Perhaps no one actually prays to idols anymore, but there are many, many "art collectors" who treat pictures, especially portraits (whether they be paintings or photographs) or statues, as though they were the human beings they portray. Without attempting to analyze or otherwise to psychologically evaluate such behavior, let us simply allude to the not unusual occurrence of people who talk sincerely to pictures or statues, and leave it at that. But what of people who treat animals (including birds and fish) as though they were human? Who (if they have not done so themselves) has not heard people speak sincerely and at length to dogs, cats, birds, fish, and what-all-else? The whole question of whether animals "understand" language we have found to be a highly volatile one (if not actually explosive). Anyone who owns (and loves) a pet "knows" that the beast "understands" what is being said; and that includes those individuals who realize that their beloved pet understands Hungarian, or whatever, and not any other language.[25] After all, wasn't Hans clever?

Now, having opened that can of worms, let us see if we can somehow prevent them from crawling all over these pages. (Re-capping the can is probably impossible because of the counterverbality of the word 'language,' with its concomitant divergent views as to who is capable of it and/or who has access to it.) With the recent growth of interest in what is frequently called "interspecies communications" [*stet!*]—what with all the experiments performed with dolphins, chimpanzees, and gorillas (these experiments are to be distinguished from any work that may have been done with parrots, mynahs, and other mocking birds)—the desire to "speak to the animals" (like King Solomon of old and Dr. Doolittle of more recent cinematic vintage) has had a renewed resurgence, especially since it was realized that although apes could not learn to talk because they do not have the physiological mechanisms necessary for vocal speech, they could be taught to sign (i.e., to communicate by means of gestures and movements) with considerable dexterity. There can, of course, be no doubt that animate beings respond to stimuli (that's what makes them "animate", isn't it?), and if response to a stimulus is what is meant by communication, then even protozoa communicate. So "interspecies communications" is, as they say, no big deal: it has been going on since time immemorial.

If, however, communication is taken to be something somewhat more complex than the simple response to a stimulus, then we must agree on the premises on which we base our arguments. If, for instance, communication is understood to mean a stimulus that B emits, in turn, as a result of hir response to a prior stimulus emitted by A, then, again, there is nothing interesting about "interspecies communications," for that, too, has been going on since the day after time immemorial. If by animal communication (zoöcommunication, perhaps?) is meant the generation of a stimulus by an animal directed to a human being with the purpose of achieving the human being's attention and response, then there still is nothing interesting about the process, no matter how it may be regarded or whatever it may be called. If, however, communication is taken to be something like a volley in tennis: I hit the ball to you, you hit it back to me, I return it, you hit it back, and so on—all done with the intention of scoring a point (or "winning the game" constituted by the activity)—then the worms start crawling all over the place. For if I speak to my dog and he or she responds by barking and I answer hir and then he or she "answers" me, and so on, who is to say what the dog's answering bark indicated in the first place—if, indeed, it indicated anything other than a persevertive action? And THERE'S the rub!

Now, this is not to say that communication is to be equated with dialogue (or its equivalent)—and, again, notice that *language* has no bearing on the process whatsoever—but it is to clarify what is implied and entailed in conceiving of the possibility of "interspecies communications." And until these premises are agreed upon and differences are resolved (or dissolved, for that matter!), anthropomorphization stands as a form that can affect nonverbal communication.

But the degree of anthropomorphization should be mentioned, too:

> In some communities, spirits, plants and stones (the latter often as representatives of spirits) may feature as addressees. Even in Westernized so-called "advanced" societies, it is common to consider pets as addressees and to anthropomorphize them, attributing to them the ability to respond. Pets can, of course, feature conveniently as addressees in a 3-participant event when for instance the wife tells her lap dog (with her husband as "audience"), "Oh, you poor little diddums—that nasty master of yours should have given you your bickies long ago." (Platt & Platt, 1975, p. 17)

It is also interesting to consider whether the "communication" that takes place between adults (other than mother, father, or other habitual caretakers) and prelinguistic infants is not affected by a factor of anthropomorphization. Observe a doting grandparent talking to an infant grandchild (even one old enough to crawl about), and come to your own conclusion.

12. *Pseudospeciation.* The taxonomic designation for modern man is *Homo sapiens,* the only extant species of the genus *Homo.* If, for some reason, certain people (or certain types of persons or certain classes of people) are considered to be "a breed apart," then such a deemnation (!) would be, according to the anthropologist, Jules Henry, an example of pseudospeciation. To consider a person "less than human" because of the way he or she looks or because of hir heritage or social background—much less to justify or to explain hir behavior on these bases—would also be examples of pseudospeciation.

This is a good term, and we shall use it to refer to "the creation of objects by communication." As Egolf and Chester (1973) explain:

> Verbal and nonverbal behaviors often contribute to the creation of objects. By treating a person as if he were inanimate, a process too often seen between parent and child, doctor and patient, clinician and client, the human, as a consequence, comes to feel and, perhaps to act, less than human. As [Elisabeth] Kübler-Ross [(1972, p. 9)] writes: "When a patient is severely ill, he is often treated like a person with no right to an opinion. . . . He slowly but surely is (beginning to be) treated like a thing. He is no longer a person."
>
> Clinically we have observed the objectification of clients. Clinicians discuss their clients, who are present, in the third person. They tell parents and spouses, "He did pretty well today," and ask, "How's he doing at home?" for example. At the same time, the clinicians' body movements

become compatible with their verbalizations as they kinesically isolate the
client who thus has imposed upon him the aura of a specimen at an auction.
We believe that this behavior, which can easily be avoided, is mortifying to
the client, who may come to view himself negatively and, in time, to present
himself in a manner which would suggest that he expects to receive such
treatment. (p. 517)

Erving Goffman (1959, p. 1) describes "classic types" and "standard
categories" of such pseudospeciated "non-persons." He cites the servant
as the classic type of nonperson in our society.

This person is expected to be present in the front region while the host
is presenting a performance of hospitality to the guests of the establishment. While in some senses the servant is part of the host's team . . . , in
certain ways he is defined by both performers and audience as someone who
isn't there.

In addition to those in such servant-like roles as elevator operators,
cabdrivers, waiters, and bartenders, Goffman says:

There are other standard categories of persons who are sometimes
treated in their presence as if they were not there: the very young, the very
old, and the sick are common examples. Further, we find today a growing
body of technical personnel—recording stenographers, broadcasting tech-
nicians, photographers, secret police, etc.—who play a technical role during
important ceremonies but not a scripted one. (p. 152)

If we may make use of Goffman's terminology (as well as his paradigm), non-persons are not ignored or disregarded, they are simply not
addressed. Non-persons make up "the backstage crew": they are, as it
were, the stagehands, the wardrobe people, and the like; and just as actors
do not perform for the backstage crew, so ordinary mortals perform for
their own particular audiences.

But this is exactly where pseudospeciation becomes manifest. Elevator operators and cabdrivers, waiters and bartenders, recording stenographers and photographers may not have scripted roles in a communication performance, but they are human nonetheless, and, as such, they are
still members of the same biological species as the performers in the activity. Like everyone else—no matter hir role—they can hear, they can
understand, and they respond—even if they remain silent while they continue with their appointed tasks, though with varying degrees of servility.

Pseudospeciators seem to be unaware of this and carry on as though
these persons—whose presence is quite evident—were simply not there.
Some of these perpetrators behave like the infants that Jean Piaget
describes to whom "out of sight" is literally "out of mind." This is frequently evident on public carriers (like buses, railroad cars, or the public
cabins of airplanes) or in public places (like waiting rooms, restaurants,

or public toilets). Such behavior on the part of ignorant, careless, or super-cilious individuals in public places may give rise to casual laughter or embarrassment on the part of the unintended auditors or observers (depending, of course, on what the perpetrators say or how they behave); but when parents talk or act in this fashion in the presence of their children, they can inflict traumas on their offspring that will remain with them as long as they live. Such institutionalized behavior is, of course, an important factor in racism (which was the point of Ralph Ellison's now classic novel, *The Invisible Man*) and other forms of social prejudice—or, in a word, pseudospeciation.

In these many ways—and unquestionably more—do human beings transact the business of encountering, dealing, and coping with each other. All this is done, in one way or other, in relation to the words and word-elements that we recognize as being the heart of human communication. Separately, and in various and varying combinations and permutations, these word-less message-carriers and impression-creators[26] repeat, emphasize, complement, contradict or serve as surrogates for verbal messages. They regulate the initiation of verbal utterances, their continuation, their interruption, and even their cessation. To refer, as we have pointed out over and again, to all these functions and activities as simply (and collectively) as "nonverbal communication" does not specify the nature, the purpose, the function, or the very structure of the particular message being considered.

We propose, therefore, that word-less message-carriers and/or impression-creators be described as follows.

1. *Paraverbal.* This term is obviously a parallel (!) to *paralanguage,* which is a term originated by the linguist, Archibald A. Hill, to describe the "music" of the voice as it accompanies the "lyrics" of the language that, together, constitute the verbal message being expressed (Trager, 1958). In like fashion, we propose the term 'paraverbal' to describe any and all body actions, bodily trappings, and other arranged circumstances that accompany what is being expressed verbally, regardless of whether the intended or actualized purpose of these accompaniments is to emphasize, to reiterate, to complement, or even to contradict what is being said in words. Thus, if I pound the table while speaking, that pounding may be described as paraverbal. If I smile, giggle, blink my eyes, or squeeze your arm while I am talking to you, those actions may all be described as paraverbal. Most physical circumstances (at least, as they have been described above), such as time, place, artifacts, and objects, if previously arranged as accompaniments to what is being said, may also be described as paraverbal. In other words, paraverbal is a term which can describe any word-less aspect of a message which is "given off" while, at the same time,

the verbal aspect of the expression is being "given" to the receiver (Goffman, 1959, pp. 4ff).

2. *Metaverbal.* The term "metacommunication" has been formulated by some writers on the analogy of *metalanguage,* which is a technical term used in the field of logic to describe the means to analyze the "ordinary language" of everyday communication. There is no way of analyzing language other than by means of language itself, and so metalanguage was contrived as a means of doing so with some degree of objectivity. The linguistic structure of metalanguage may be exactly the same as that of the object language it is meant to analyze: the vocabulary and the rules of syntax of the two may be identical. But metalanguage can talk about NOTHING OTHER THAN the object language it is meant to analyze. Phonetics is an excellent example of metalanguage. One cannot SAY ANYTHING IN phonetics (if one could, phonetics would then be a form of ordinary language), but phonetics does permit one to say a great deal ABOUT ordinary language—certainly much more than could be said about the sound of ordinary language (which is, of course, the province of phonetics) than could be said in ordinary language itself.[27]

Metalanguage has come to be defined as "language used to talk about language." This is a good enough definition, though it is rather loosely stated. When, however, "metacommunication" is defined as "communication about communication," the looseness of statement mitigates any value the definition might otherwise have. It is better to define metacommunication as communication which conveys something about the communicator OR the communication OTHER THAN is contained in the communication itself. Suppose I tell you what I think of the present state of the world, and suppose that something about my telling it conveys the impression that I know (or do not know) what I am talking about, or that I am someone to be heeded (or disregarded) on the subject of the present state of the world, or that my views on the subject are pithy (or frothy), etc., etc. Whatever it may have been that gave you any one (or all) of these impressions may, then, be identified as being metacommunicative. You were, in a sense, told something BEYOND what was actually communicated: hence, METAcommunication. On that analogy, we offer the term 'metaverbal' to describe the word-less aspect of a message that communicates the transmitter's personal affect in regard to the message he or she is transmitting verbally or otherwise. Thus, one's reluctance to tell someone bad news will be demonstrated metaverbally. The narration of a witnessed event will demonstrate the narrator's feelings about the event metaverbally. The intensity of a paraverbal gesture used to emphasize a verbal statement will be demonstrated metaverbally: I may pound the table paraverbally to emphasize the words I am saying, but the intensity of that pounding will communicate my feelings metaverbally.

3. *Homoverbal.* The word 'gesture' usually refers to a communicative movement of some part of the body, such as a nod of the head, a shrug of the shoulder, a wave of the arm, etc. But we use the word more broadly here (not limiting its reference to kinesics) when we say that 'homoverbal' refers to any word-less gesture, no matter how it is performed or in what channel, which can be directly and entirely translated into words. Such body actions, bodily trappings, or arranged circumstances are, then, the same as words from the point of view of communication, and so should, logically, be referred to as being homoverbal, which means "the same as words." Homoverbal cues would obviously include what Ekman and Friesen (1969) call "emblems." As they phrase it:

> Emblems are those nonverbal acts which have a direct verbal translation, or dictionary definition, usually consisting of a word or two, or perhaps a phrase. . . . An emblem may repeat, substitute, or contradict some part of the concomitant verbal behavior; a crucial question in detecting an emblem is whether it could be replaced with a word or two without changing the information conveyed. . . .
>
> Emblems occur most frequently where verbal exchange is prevented by noise, external circumstances (e.g., while watching a play), distance (between hunters), by agreement (in the game of charades), or by organic impairment (the deaf mute). In such instances, emblematic exchange carries the bulk of messages which would typically be communicated through words. Emblems, of course, also occur during verbal exchange. . . .
>
> The tracing of a body outline of a woman is an iconic-pictorial emblem in which the hands draw a picture of a shapely woman to state sexual attractiveness. The making of a fist, or shaking of a fist, is usually an iconic-kinetic emblem.[26] (p. 57)

The hitchhiker's thumb and the extended index finger are both common, easily recognized emblems, or, to use the terminology being proposed here, homoverbal gestures. Obscene gestures, which are obviously homoverbal, for the most part, manage to be shown even in polite society, despite suppression of their verbal glosses, and regardless of the intention or awareness of the sender. Who can forget the picture (shown on the front pages of many newspapers across the country) of Vice President Nelson Rockefeller responding to a hostile audience homoverbally by "giving them the finger"?

4. *Heteroverbal.* In the attempt to fulfill the various purposes of communication, there are instances when it is more effective, much less more expeditious, to transmit a message by some means other than words, even though the information could be encoded verbally. Thus, pictorial, sche-

matic, analogic, or mimetic representations all could be put into words if they had to, but they serve communication far better heteroverbally. If the sender feels that it would not meet hir needs to do so, he or she may choose to communicate a particular statement by some other means, that is, heteroverbally. If I am telling of an incident in which the strange gait of one of the participants is crucial to the point of the story, I would do better if I demonstrated the gait mimetically (" . . . and he walked over to me like this . . .") rather than by trying to describe it in words. A witness could give a good description of the perpetrator of a crime in words, but how much better if he or she could draw a picture or put together a composite photograph of the perpetrator heteroverbally. If you received a note from someone that said something like "Please see me at once," you would very likely respond to the verbal part of the message as you usually would, all other things (as they say) being accounted for. But if that same verbal message were to be scrawled hurriedly, in oversized script perhaps, or in colored ink, or heavily underlined, or followed by an exclamation mark, the dot at the bottom of which almost punched through the paper, then the graphemics of the message would be transmitting "URGENCY!" heteroverbally. This is the sort of thing (among others, we can be sure) that Kendon must have had in mind as indicative of the nonverbal aspect of "the written word."[29]

Heretofore, it would have been virtually automatic to explain the meaning of the phrase "to communicate by means other than by words" simply by saying that to do so was "to communicate nonverbally." If one wanted to be more rigorously specific in explanation, one might say that instead of phrasing something verbally, one did so kinesically. Though both of these characterizations are true, it should be apparent that describing the transmission of a message by means other than by words as heteroverbal has the obvious merit of distinguishing the function and the purpose of the action from any other that might also fit under "the nonverbal umbrella."

One could quibble about the distinction (if there is any) between homoverbal and heteroverbal. After all, a traffic cop at a busy intersection, for instance, could not do hir job effectively by vocally urging motorists to stop or go. That would simply not be practical. Then are the traffic cop's hand signals homoverbal or heteroverbal? Anyone who bothered to ask a question of this sort with any degree of seriousness would obviously have missed the point. In the first place, none of the categories offered here is discrete or exclusive. In the second place, these terms are meant simply to be more efficacious means of description rather than as labels of identification. And in the third place, such a question would be indicative of a failure to understand the meaning of 'heteroverbal' truly in the first place. A traffic cop's hand signals are unquestionably homoverbal because they can be directly and entirely translated into words without changing the information conveyed. 'Heteroverbal,' on the other hand, refers to those instances for which words can be found, but their number could not be restricted to "one or two, or perhaps a phrase" in transliteration. If a verbal description were attempted in these instances, it would be clumsy, difficult, turgid, repetitive, and what-all-else. That is why tracing arcs in the air with fingers stiff and close together and palms flat to visualize aerial loops and dives, for instance, was chosen by the sender instead.

Two anecdotes, both involving the philosopher Ludwig Wittgenstein, will further illustrate the nature of what is heteroverbal. Once, when asked how tall he was, Wittgenstein stood up, with his hand flat on the top of his head, and said, "I'm this tall." Innocuous as this may at first appear, it illustrates dramatically the difference between a verbal and a heteroverbal response. One would normally have expected the answer to have been something like "five feet, ten inches" (or whatever). Being "this tall," however, is just as true heteroverbally (it might, in fact, be even more true!) than any verbal response to the query.

The other anecdote is best presented in the words of the writer from whom we learned of it:

> According to [Wittgenstein's] *Tractatus* [*Logico-Philosophicus*], language is a picture of reality: language depicts the logical structure of facts. Wittgenstein's repudiation of this view is one characteristic difference between his earlier and later work, between the Tractatus and the Philosophical Investigations. Naturally there was a period of transition when Wittgenstein was moving away from the ideas in the Tractatus before decisively rejecting them, a period when the ideas in the Philosophical Investigations were beginning to take shape.... The story of how Wittgenstein came to doubt and then reject the picture theory is this. A Cambridge colleague was the Italian economist Piero Sraffa, with whom Wittgenstein often discussed philosophy. One day when Wittgenstein was defending his view that a proposition has the same logical form as the fact it depicts, Sraffa made a gesture used by Neapolitans to express contempt and asked Wittgenstein what the logical form of that was. According to Wittgenstein's own recollection, it was this question which made him realize that his belief that a fact could have a logical form was untenable. (Hartnack, 1962, p. 62).

This anecdote may be apocryphal, but it serves to illustrate that the gesture Neapolitans use to express contempt is heteroverbal. As such, the gesture is communicable and meaningful, but it cannot be put into verbal form, and certainly not into words that would be similar in form or even represent similar forms. If the logical form of a depicted fact must be put into verbal form—and that does not necessarily mean that it must be put into linguistic form: it could be stated numerically, or in some other symbolic form; but however it might be done such a formulation would still be VERBAL[30]—if the logical form of a fact must, then, be verbal, then that verbal formulation has no similarity—neither in sound, in sense, in shape, nor in style—to that which is formulated in some other way than by words, or that (by morphology, if not by definition) which is HETERO-VERBAL! Sraffa's gesture may have been the straw that broke Wittgenstein's refusal to repudiate his previous philosophical views, but it may have done so for a reason that was beside the point.[31]

5. *Exoverbal.* As Kendon pointed out, "communication in face-to-face encounters, at least, is a complex phenomenon in which all aspects of action are involved."[32] Some of these pertain to the message being transmitted, but some pertain only to the sender: indicating hir presence and hir state of being at the time of transmission. Since, in most instances, it is difficult (if, indeed, it is at all possible) to separate one from the other, it would be helpful if this distinction could be made when it is necessary to do so for purposes of analysis and understanding. Just as the human being is capable of producing all sorts of noises, many of which are not used in either verbal or vocal communication, and do not occur (at least as far as we know) in any natural language spoken anywhere in the world, so, also, is the human being capable of producing all sorts of movements and actions that are not part of any message system that may anywhere be in effect at any time. Such sounds, movements, and actions, which are, as it were, outside of the realm of words, may, then, be described as being *exoverbal.* Thus, for instance, hiccoughing and borborygmus (stomach rumbling), handclapping and finger-snapping, sneezing and honking, lip-smacking, slurping, teeth-sucking, and lots more, are all noises produced by human beings frequently enough to be easily recognized as being "outside" the realm of linguistic morphology and word-construction; so also are the body tremors, shudders, tics, and such other body movements and motions as are recognized as being "outside" the realm of what Birdwhistell has called "body motion communication." If such sounds, movements, and actions indicate anything, they indicate the presence of the sender and hir present state of being. And so, though such sounds and actions may be characterized as being indexical of the sender, they are exoverbal as far as their being part of any message that may be transmitted. Some, if not all, of these actions would be classified by Ekman and Friesen (1969) as "adaptors," or behaviors designed to satisfy the physical or emotional needs of the sender.[33]

6. *Averbal.* If something were to be truly nonverbal, it would have to be something that either could not be put into words or something for which there were no words. Words (or at least verbalization) would simply not be involved, or would not pertain, or would be beyond anyone's ability

to achieve it. Wittgenstein must have had something of this sort in mind when he spoke of that "whereof one cannot speak." One could not speak thereof because there would be no words that could represent, explain, or even describe whatever it is whereof one cannot speak. Nor could there be words therefor. Any words that might be specifically created therefor might paint a lovely verbal picture, but neither the picture nor the words would have anything to do with what they were supposed to represent. Now, if this statement sounds reminiscent of Meister Eckhart, the fourteenth-century mystic (who was supposed to have once said, "Whatever you say God is, that He is not!"), we do not gainsay it.[34]

Averbality, however, should not be confused with heteroverbality. For instance, when writing on the subject of olfactics, Ruth Winter (1976, p. 146) said, "No odor can be described verbally in English [or in any other language, for that matter] in such a way that can be immediately recognized or summoned up." The reason why no odor can be described verbally is because of the heteroverbality of the sense of smell. How does one describe—that is, verbalize—the scent of roses? How does one differentiate it from the scent of lilacs? The fact is that one cannot. But that does not make scent averbal. Scents can be artificially or synthetically produced.[35] These artificial or synthetic stimulants can specifically evoke the recognition of their referents even though they may do so in their own particular way in their own particular channel. Hence, scents and odors are heteroverbal rather than averbal.

Contrast this with consciousness. "Consciousness cannot be defined; we know what consciousness is but we cannot provide an unequivocal definition for others of what we ourselves are able to grasp clearly." (Guiraud, 1957, pp. 10–11) Consciousness, then, is averbal. We all KNOW what it is, but we can neither define it nor describe it—because it is averbal! But (says a voice from the balcony) we can talk about consciousness! What else are we doing now? Does that not, then, make it verbal? No, it does not! Because all that the word 'consciousness' does is to NAME it! "Under the entry 'consciousness,'" Guiraud (p. 11) tells us, "Lalande's *Vocabulaire de la philosophie* [goes on to say merely that] . . . 'We become less and less conscious as we gradually fall asleep . . . and we become more and more conscious when a noise wakes us gently.'" What, then, does this verbal description tell us in and of itself? Absolutely nothing! Reread it, and realize that the VERBAL DESCRIPTION SAYS NOTHING. It is meaningful (after a fashion) only because we have all gradually fallen asleep and because we have all gently awakened, and so we know what those conditions feel like. All that the verbal description is, in this instance, is a kind of tautology, a repetition of what it is presumed is already known.

But what if it is not already known? What happens then? Well, suppose, for a moment, that you did not know what a zebra was, nor had you ever encountered a zebra without knowing that what you had encountered was a zebra. In such circumstances, a zebra could quite easily be described to you verbally. But suppose you did not know what an orgasm was, nor had you ever had an orgasm without knowing that what you had had was an orgasm. Can you imagine describing an orgasm verbally to someone in such circumstances?[36] An imaginative, articulate, literate, sensible, rational, well-meaning, fluent individual might paint a pretty word picture of an orgasm. Many well-known writers have: Ernest Hemingway's description in *For Whom The Bell Tolls* comes immediately to mind. A beautiful description! But that is not what an orgasm is! And the same is true in the case of *friendship, hatred, grief,* and the chest pains of angina pectoris. Meister Eckhart might have had the right idea all along!

We live in two worlds: the averbal, and the verbal. Everything that we experience is averbal. And what do we talk about mostly? Our experiences, of course! (Beware the counterverbality of the word 'experiences' here, and be sure to stick to the meaning we started out with: that everything WE EXPERIENCE is averbal; hence, we mean 'experience' here as sensation, feeling, and physical and emotional response.) But in order to talk at all, we must verbalize. (As T. S. Eliot's Sweeney Agonistes said, "I gotta use words when I talk to ya!") So we try to verbalize our averbal experiences as best we can. Those of us who succeed beyond a babble (perhaps THAT is why so many people who are unsuccessful in this regard reiterate "Y'know! Y'know?" so frequently!)—but those of us who do manage somehow have developed some ingenious ways of trying to verbalize the averbal. When, in our efforts to communicate, words fail us, we supplement them, complement them, modify them, elaborate them, repeat them, and do whatever else we can do. It is a shame that we are understood so infrequently. But, then, that is quite another story!

One would conclude that that which is averbal is the stuff of which Wittgenstein warned we must be silent. That may be so as far as any verbal examination, any verbal expression, or any verbal exchange may go, but, still, all of us have, at one time or other, been involved in some situation in which our correspondent's eyes have lit up in recognition of what we are referring to (but were unable to express or explain—because of its averbal nature). "Do you know what I mean?" we plead, hopefully (if not with a measure of desperation). And if you do, we are blissful! We got across to you! You understood! You must be like us, because you think and feel as we do! You're our kind of person! And we love you! (This may well be the basis for the recognition of a sympathetic soul in the face of a stranger across a crowded room. This may, indeed, be the basis for the folklore of "love at first sight.") Though not a word need be said—(nothing, in fact, CAN be said—because the matter is averbal, and so CAN NOT be said)—and yet . . . you understand! What better evidence of extrasensory perception (ESP)?[37] Y'know? Perhaps, then, THIS is why so many people ask whether "y'know" so often and so repeatedly. Y'know?

But the trouble here is that it is virtually impossible to differentiate the averbal from the dysverbal. The averbal cannot be spoken, cannot even be spoken about (beyond naming it, or making some other reference to the fact that that sort of thing exists). But that which, though verbal, is difficult to phrase, or difficult to express, or difficult to explain—that which, though verbal, we do not have the words for, or we do not know the words for[38]—that leaves us just as speechless as that which is truly averbal would. In which case, what better way to mask our inability, or our ignorance, than to simulate a valiant effort to cope with the averbal? Y'know? Y'understand what I mean?

If, as may well be the case, we really do not know what is going on, if we are not sure of ourselves in the situation going on at the moment, if we are a leetle too empathetic of the poor devil who is trying to talk to us (who, but for the grace of G-d, could be us!), we assent: yes, yes, we know! we understand what you mean! we get it!

GET WHAT?

The fact is (more than likely) that the guy[39] who asked us in the first place if we knew, if we understood, if we got it (y'know?), DIDN'T KNOW, DIDN'T UNDERSTAND, DIDN'T GET IT!!! For IF he or she did, HE or SHE COULD HAVE SAID WHAT HE OR SHE MEANT (and so have avoided the whole mess from the beginning).

Now, saying what one means does not necessarily entail "putting it into words" (IT may not BE words, in the first place). But saying what one means does imply that you can SAY that you do not know the words, or that you cannot find the words for whatever it is that you may have been talking about (or have been wanting to talk about). In which case it might be the better part of valor TO SHUT UP!

So much, then, for averbality, and its possible relation to verbalization.

If we were now to review Kendon's criticism of Key's bibliography of nonverbal communication that we discussed earlier, and if we were to do so in the vocabulary that we have set forth here, we would find that in circumstances of any kind in which there were words involved—no matter how they were transmitted, whether in speech, in writing, in African drum language or Canary Island whistle language, in Amerindian smoke signals or Boy Scout flag semaphore—in short, if the message is in words, it would be *verbal*.

If, while transmitting words (or elements of them), we were to accompany them with, as Strawson says, "a gesture or a drawing or the moving or disposing of objects in a certain way," then such communication would be *paraverbal* (alongside what the words are saying).

If, while transmitting words, the sender's presence, or hir feelings, beliefs, or attitudes about hirself, the receiver, the contents of the message, or the state of the world at large, were to be made manifest in the transmission of the message, then that aspect of the communication would be *metaverbal* (beyond what the words are saying).

If circumstances warrant, justify, or require that some means of transmission be used other than speech or writing, but such means can be directly and entirely transliterated into words without changing the information conveyed, then such communication would be *homoverbal* (the same as what the words are saying).

If circumstances warrant, justify, or require that some means of transmission be used, such as pictorial, schematic, analogic, or mimetic representation, because a part or parts of the contents of the message do not lend themselves to efficient encoding in words, then that kind of communication would be *heteroverbal* (another mode in which to communicate more efficiently what the words would be saying—if they could).

If sounds or actions are outside the realm of words, but are nonetheless seen, heard, felt, or otherwise sensed in the communicative situation, then that aspect of the overall signal being transmitted would be *exoverbal* (outside the realm of words).

And, finally, if we are aware of feelings or sensations which could not be put into words because of their ineffability or because there are no words for them, then such aspects of our awareness would be *averbal* (there are no words).

Kendon may think that "'non-verbal communication' is a non-topic," but the word 'nonverbal' is not to be thought of as a "non-word"—not even after the reclassification that has been presented here. Finding names for specifics does not mean genocide for generals (or verbocide for words!). 'Nonverbal' is a perfectly good word to refer in a general sense to that which does not say something in words or by means of words. When, however, it is necessary to express a specific and exact wording in reference to the process of communication by some means other than by words, we now have the vocabulary with which to do so.

NOTES

1. A word about typography, orthographical diacritics, and other such graphic conventions as they are used in this essay: Items enclosed in single quotation marks, like 'hunger,' represent verbal signifiers, or words, names, labels, and

the like. Items in italics (or underlining in typescript) represent the phenomenon which is signified, as in *hunger*; and this would include foreign terms, foreign names, and statements in foreign languages, as well as the titles of publications. Items enclosed in double quotation marks, as in "hunger," represent verbal signifiers or words that are rhetorically modified in their specific use. Thus, to speak of an intellectual "hunger" would be a rhetorical extension of the word 'hunger,' which is the name of the physiological phenomenon of *hunger*. Since someone's use of words is a rhetorical extension of those words, quotations will, as is usual, be enclosed in double quotation marks. That leaves emphasis. Though usually indicated by italics (or underlining in typescript), emphasis is indicated here by SMALL CAPITAL letters.

We are very much aware that these modifications will take some (little) getting used to; but we must suffer the consequences of any inconvenience this may cause our readers—such as it may be for those readers to whom it may pertain. As a would-be healer of orthographical diseases, we must (as should all physicians) try to heal ourself first. Typography, punctuation, and other diacritical markings, as well as all the rest of the formal conventions of writing and printing, can full well give rise to counterverbality in, of, and by themselves. We sincerely hope, therefore, that this caution will prove to be salubrious.

2. Frankly, we do not know who Paul Anderson is—or was, for that matter—but the quote heads Chapter 3 in Katz (1972, p. 56) and it certainly is apropos here.

3. In a dissertation such as this one, verbalists like us must always look to their laurels, but we do retain the privilege of resting on them as well. In our view, then, 'oral,' like 'anal,' connotes a body orifice provided by nature for reasons more fundamental (!) than verbalization. We would, then, rather talk about words as being 'spoken,' or as being 'vocal,' when they are not 'written' or 'printed.'

4. Be it noted that we do not mean to deliberately misinterpret Leach here: he does go on to explain that it is the patterning of the sound that is what makes the utterance verbal. The statement, however, still imputes the identification of verbality with vocality. In fact, Leach's statement on the very next page of the same essay (141) to the effect that "writing a letter is a non-verbal kind of behavior, but it conveys information by verbal means," which is intended to illustrate the fact that "there is no straightforward distinction between verbal and non-verbal communication," rather than clarifying it, only seems to obfuscate the observation still further.

5. In *Early Language* deVilliers and deVilliers (1979) apologize for having "used the male pronoun generically because it seemed awkward to do otherwise" (p. 2n). Well, dear reader, be advised that we here are going to do the otherwise—even at the risk of further provoking the annoyance of those of you who may still be with us (cf. note 1 *supra*)—not out of some sort of inherent perversity, but simply because, as Milt Horowitz used to say, "there comes a time in every man's life when he must set aside his principles, and do what's right!" And so, to avoid any deliberate charge of sexism, we have adopted the conventional forms of the pronouns and adjectives "he or she,"

"him or her," "himself or herself," "hir" (a composite of 'his' and 'her'), and "hirself" ('himself'/'herself'), and we shall use them throughout this essay. The so-called "feminist movement" has moved us (in this regard, at least!), and for sure!

6. A word about the orthography of "the verbal term 'nonverbal,'" which is, after all, what this whole megillah is all about. Common sense (!)—perhaps better (even if somewhat pleonastically) a native *Sprachgefühl*—would incline one to assume a difference of some kind between "nonverbal" and "non-verbal." Well, let us make it perfectly clear at this point that we make no distinction, we intend no distinction, and we mean to imply no distinction between the two. Different writers do spell the term differently. Perhaps THEY intend a distinction: we do not intend to probe, and certainly not to argue, the point. We simply quote them. If they hyphenate, we do when they do. But when we use the term, even when we refer to their use of the term, we do not hyphenate it.

7. In the September 1978 issue of *Asha* there is a full-page advertisement on the back cover which urges the readers of the journal to "open a new world for the non-vocal/non-verbal" with "the voice that actually talks for them!" The advertisement (which was run again in several succeeding issues of the journal) tells us that "until now, communication aids could only communicate in symbols, in pictures or by spelling words," but this device, which is "a hand-held electronic voice synthesizer which can produce virtually any word in the English language," is "a remarkable communication breakthrough" because it "actually talks." Such a device, the ad goes on to say,

> opens a new world of communication for children and adults who do not have oral communication abilities. This includes people who have cerebral palsy, aphasia, or who have had strokes, spinal cord injuries, brain damage, or other afflictions or birth defects which have left them non-oral.
>
> For a wide range of orally-disabled, two ... models are available: [one] offers a touch-sensitive pre-programmed keyboard with 473 sounds, words and phrases. [The other] works like a calculator and has 991 sounds, words and phrases in its program. Each is accessed through a three-digit code.
>
> Both [models] are programmed with phonemes and morphemes. This enables the user to create a virtually unlimited English language vocabulary by combining basic component sounds.
>
> Now [speech–language pathologists] can open a new world for the non-vocal/non-verbal.

This is a commercial advertisement placed by a manufacturer whose purpose it is to sell his or her product. One would assume that manufacturers who advertise in professional journals would make it their business to address their prospective professional clients (in this case they would be speech–language–hearing pathologists) in an appropriate professional way, using the currently accepted technical terminology (or at least that which is current among practising professionals at the time the advertisement is published). This is why we quote this advertisement at such length here. If this advertisement is indicative of the current use of the terminology among practising speech–language–hearing pathologists, particularly with regard to the use of the term "nonverbal," then we rest our case.

It is certainly not our intention either to dilate this already distended note or to hold forth any further on the polysemia, much less the multiple usage, of the term 'verbal.' But we cannot desist here without making some reference to that not infrequently heard usage which takes 'verbal' to mean an excessive use of words in vocal speech. A person will be said to be "very verbal" when what is meant is something akin to 'garrulous,' 'voluble,' or 'verbose.' This usage would remain an incidental item of dialectological trivia but for what it does to the converse term. A "nonverbal person" would, then, be dumb, mute, or speechless! O, how many times must we cry, "O, mores!"?

8. We should note the recent appearance in print of the phrase "precise words" which are described as antidotes to the fuzzballs of jargon (which are also subsumed by what we are here descrying [no, that is not a misprint!] as counterwords).

9. Anyone curious enough to do so may glance at the literature on general semantics—not to mention General Semantics (which is certainly not the same thing)—as well as what has been written on the other specific types of the stuff, all of which have to do, more or less, in one way or another, as the case may be, with *meaning*.

We cannot go into it fully here (it would take an essay at least as long as this one to treat the subject adequately), but this sort of thing can also be the consequence of historical circumstances of such a nature as to preclude attribution to any one person, group, movement, or circumstance. Consider the polysemanticity (now, that is an impressive sesquipedalianism, but it is as euphemistic a bit of jargon as can anywhere be found to blanket the meaning of what one is trying to express!)—consider the polysemanticity of the English word 'love,' which in and by itself comprehends the separate concepts delineated in Greek as *agape, eros,* AND *philia.* And the English word 'language' which in and by itself comprehends the separate concepts delineated in French as *le langage, la langue,* AND *la parole.* "Polysemantemes" of this sort should not be confused with (and so dismissed as) homonyms, such as the word 'glasses.' "Containers to be drunk out of" are not the same as "spectacles to be peered through" (though both are frequently made of glass, and so go by the same name). The name given these two classes of objects is a word whose meaning may be said to have "expanded" semantically; hence, that "word" (actually, there are TWO words here, two homonyms, both of which are homographic (spelled the same) AND homophonic (pronounced the same) but which have different meanings)—hence, the "word" 'glasses' may be said to be polysemantic. But the English language has, and has had, only the one word 'love.' When it was learned by speakers of English that speakers of ancient Greek recognized and separately named

(1) *agape*: religious (specifically Christian) devotion and transcendence, which is something completely different from (and having no relation to) (2) *eros*: sexual attraction and erotic attention, each of which was completely different from (and had no relation to) (3) *philia*: fraternal responsibility and concern, all three were, and are, popularly conceived as "having a common denominator," which is a classic example of rationalization after the fact. Much the same circumstances have prevailed in English in the case of the word 'language,' as was indicated so pointedly by Saussure (1959).

10. It is apropos at this point to mention New York State's "Plain English" Law, which was signed by Governor Hugh Carey on 5 August 1977 as L.1977, ch.747, and was further amended in May of the following year as L.1978, ch.199. The new Sec. 5–702 of the law requires that

> each written agreement entered into after November, 1978, for a residential lease or for money, property or services for personal, family or household purposes involving less than $50,000 must be ... written in a clear and coherent manner using words with common and everyday meanings.

Writing in the *New York State Bar Journal*, Richard A. Givens (1978) says:

> The 1977 Act grew out of concern over consumer credit contracts couched in unintelligible language, understandable only by an attorney versed in the arcane subtleties of creditor remedies, or in what ... has [been] called ... 'bafflegab.' (p. 480)

The intention of the statute, it cannot be gainsaid, was noble, but, we daresay, it was linguistically and communicatively naive. Legal bafflegab may be couched in arcane subtleties, but at least those subtleties have been fought out in the courts and so are legally defined (even if understood only by grasping legal-eagles). It is too early to tell at the time of this writing, but the likelihood is that as a result of the "Plain English" in those consumer credit contracts, there will be even more litigation (and, hence, more business for those legal-eagles) than before—simply because of the ubiquity of counterverbality, especially in "'plain' English"!

11. Although the point here is that despite Kendon's excoriation of Key's profligacy in her recognition of so much as being relevant (in the broadest possible sense of relevance, of course) "to understanding how people manage to communicate with one another when co-present," he still censures her failure to include yet another area of consideration, namely, that of communication by way of the written word, one may wonder nonetheless what there could be about the written word that is NONVERBAL. The verbality of the written word is so obvious as to be tautological, but what could be nonverbal about writing? Well, dear reader, have no fear! This will all be explained in due time—at the appropriate time—later in this essay: as the pedants would have it, *q.v.infra*.

12. A taxonomy that would categorize "a non-avian zoology" would certainly recognize "a zoology of the hairy quadrupeds"!

13. Oscar Wilde once said that he could resist anything but temptation. Well, we can appreciate that, because we cannot resist the temptation at this point of noting Barnett Newman's remark to the effect that although it may be for the critics what the Holy Bible is for the fundamentalists, esthetics for the artists is like ornithology "is" for the birds.

14. The problem is not unique. Korzybski ran into it early on in his career, perhaps even prior to the time his *Science and Sanity: An Introduction to Non-Aristotelian Systems and General Semantics* was first published by the International Non-Aristotelian Library Publishing Company of Lancaster, Pennsylvania in 1933. Korzybski spent the rest of his life (and some of his "disciples" and apologists are still) explaining that what he meant by "non-Aristotelian" was not to be construed as ANTI-Aristotelian.

15. In a way, the description of dialects as "nonstandard" has suffered a somewhat similar fate.

16. In the 1976 translation by D. F. Pears and B. F. McGuiness, Wittgenstein's famous koan is rendered, on p. 3, as "what can be said at all can be said clearly, and what we cannot talk about we must pass over in silence." We take the liberty here of rendering what, to our taste, is at least a more alliterative version of the same statement.

17. We considered the possibility of traducement in the translation, but the original German says THE SAME THING: *"Was sich überhaupt sagen lässt, lässt sich klar sagen; und wovon man nicht reden kann, darüber muss man schweigen." Sagen* means "to say," *reden* means "to speak," and *schweigen* means (perhaps most closely in English) "to shut up" or "not to speak." The problem—again, it is a matter of counterverbality—inheres in the referential implication of *sagen* in German and 'say' in English. Both terms comprehend BOTH generalized "expression" (regardless of mode) as well as specialized "vocalization."

18. This is demonstrable in the restriction on the utterance of *The Ineffable Name* of the Deity—particularly in tetragrammatal form—by those who follow the practices of Orthodox Judaism. Even THE SIGN of *The Name* in written form (which is, after all, silent) will be avoided by the truly devout, who will not even spell out (much less utter) the word for "G-d."

19. Anyone curious as to why Korzybski chose to call the system he created "General Semantics" (with an upper-case 'G' and an upper-case 'S' to distinguish it from 'general semantics') can find out why in an article I wrote (Newman, 1961), entitled "General Semantics and Academic Phagocytosis."

20. Gray and Wise do not mention "secondary sex characteristics," but biologists (maybe even sociobiologists) still speak of them, do they not?

21. Compare Ray Birdwhistell's (1970, p. 3) colorful comment: "A human being is not a black box with one orifice for emitting a chunk of stuff called 'communication' and another for receiving it."

22. To illustrate further that (as Letitia Raubicheck used to say) "the whole person speaks," The *New York Times* once ran a story about a well known (and highly successful) movie actor who was embroiled in litigation with his pro-

ducers over who had the rights to the films he had starred in. The *Times* reported that in an interview "the actor sat in his lawyer's office close to tears. He tried to give an unemotional account of the seizure of the films he had acted in. But his arms flailed the air as the words were not themselves adequate." Even a highly skilled actor, who should have had more experience than the average person in the practice of deliberate, overt expression, can find that words may not be adequate by themselves to express what is meant, much less what is felt.

23. Even if we were to concede the existence of true anatomical hermaphrodites (it should be obvious that we are not at all inclined to do so), their occurrence would be so rare as not to affect our discussion here.

24. We look forward to Edgar A. Gregersen's forthcoming book on sexual practices around the world in which there is to be a whole chapter on genital jewelry. (He is Professor of Anthropology at Queens College of the City University of New York.)

25. For those who do not get this reference, or who would like to know more about it, may we suggest Paul Watzlawick's (1977) *How Real is Real?: Confusion, Disinformation, Communication.*

26. The interested reader will find Erving Goffman's fuller treatment of this aspect of the process in the book by him that has already been cited, as well as in others of his works.

27. Again, should any reader care to pursue the matter entailed here, may we suggest that he or she look into Adam Schaff, *Introduction to Semantics* (1962), particularly chapter 2.

28. Anyone familiar with the literature of nonverbal communication may have been wondering when we were going to get around to acknowledging and referring to this item—precisely because of its apparent similarity to this essay. Well, here it is: we acknowledge Ekman and Friesen's article and we are here referring to it. We would like to make it perfectly clear, however, that we have not been suppressing acknowledgment and reference until we could no longer do so. This essay is NOT the same as Ekman and Friesen's, although, since we are writing on the same subject, we inevitably touch the same bases (we may not be playing in the same ball park, however—much less in the same league!). When we happen to touch the same base-pad, as is the case here, we admit it. That's all there is, and, as far as we are concerned, there ain't any more!

29. Perhaps it is too small a regard for those readers who have paid almost as much attention to the notes thus far as they have (hopefully!(!)) paid to the text, but THIS is the *q.v.infra* referred to in note 11 above.

30. We full well realize the weighty implications inherent in this conception of the meaning of *verbal*, but since the arguments therefor are not involved in the point being made here, we shall leave it be.

31. Before leaving this subject and going on to the next (while, at the same time, not trying to make any more of it than is necessary to resolve and explain possible relationships and involvements), we do want to acknowledge our awareness of the possible similarity that might be noticed in what we have

designated as heteroverbal and what Ekman and Friesen (1969) call "illus-trators" and the various types thereof. Again, we recognize that we are all running on the same basepaths here, but, as we have said, we are not playing the same game.

32. See note 10, *supra*.
33. Ekman and Friesen do not consider sound at all.
34. Anyone of a scholastic bent of mind (medieval, that is!) can find a great deal of this stuff set forth in Blakney (1941) and Ancelet-Hustache (1957).
35. This is the sort of thing Aldous Huxley wrote about in *Brave New World* when he described the movies of the future. Huxley wrote about "feelies" rather than "smellies," of course, but several years ago in New York City an enterprising entrepreneur did present motion pictures that projected odors throughout the theater. The reviews said that the experiment was successful. But it did not last—because the odors did.
36. We can dare say that that is one of the reasons why parents (and other older individuals) always put children off with "Wait till you grow up!" Unfortunately, not everyone does.
37. We should be satisfied with the assumption that it is not necessary to do so here, but nonetheless we want to make it perfectly clear that this rhetorical question is NOT to be construed as indicative of our belief in, or our advocacy for a belief in, extrasensory perception. Let us repeat (for the purpose of making it perfectly clear—again) that what we are trying to show here is why great numbers of people rationalize their resort to such a justification for this belief. There are similar rationalized justifications for resort to belief in the supposed predictive powers of astrology—but we neither believe in that nor do we advocate belief in that either.
38. We have met face-to-face the situation of verbality for which one does not have the words. It happened across languages, as it were. We do not claim complete mastery of the German language, but we do feel that we can man-age the language receptively almost to the point of satisfying our needs and requirements. There are times, however, when we run into an opaque term or an arcane construction, and we cannot get on with what we are doing. So we usually take the easy way out and ask someone who knows. Now, we hap-pen to have several friends whose native language was German, and who were educated in Germany up to the university level, at which time they emigrated to the United States. Their knowledge of German, one would assume, would be the equivalent of our knowledge of English. When we asked—it was the first time we did so, and so we asked with considerable humility—what (the German word in question happened to have been) *Mundartforschung* meant, believe it or not, they could not tell us! They fum-bled and babbled and apologized rather clumsily for the difficulties encoun-tered in trying to translate from one language to another; they murmured something having to do with the difference in semantic fields in different languages; they—finally, we realized that they simply did not know the word! "It is some kind of research"—*Forschung* means 'research'—was the best they could do. So we had to undertake further bibliographical research on

our own to find out that *Mundart* (literally, "mouth art") meant 'dialect' and *Mundartforschung* meant 'dialectology'. Don't laugh! Do YOU know what 'fanon' means? And that is English! And do you know what 'maniple' means? But, enough!

39. Since we have lapsed into the mother lode of slang, we should explain that we here use the term 'guy' generically, not genderically (!). We foreswore sexual chauvinism in our language use early on in this essay, and we mean to abide by our commitment.

REFERENCES

Ancelet-Hustache, J. *Master Eckhart and the Rhineland mystics.* New York: Harper & Row, 1957.

Asha, 1978, *20*, back cover.

Birdwhistell, R. *Kinesics and context.* Philadelphia: University of Pennsylvania Press, 1970.

Black, M. *Language and philosophy.* Ithaca, N.Y.: Cornell University Press, 1949.

Blakney, R. *Meister Eckhart.* New York: Harper and Row, 1941.

Bloomfield, L. *Language.* New York: Holt, Rinehart & Winston, 1933.

Burgoon, J. K., & Saine, T. *The unspoken dialogue: An introduction to nonverbal communication.* Boston: Houghton Mifflin, 1978.

de Villiers, P. A., & de Villiers, J. G. *Early language.* Cambridge, Mass.: Harvard University Press, 1979.

Egolf, D. B., & Chester, S. L. Nonverbal communication and the disorders of speech and language. *Asha*, 1973, *15*, 511–518.

Ekman, P., & Freeman, W. V. The repertoire of nonverbal behavior: Categories, origins, usage, and coding. *Semiotica*, 1969, *1*, 49–98.

Farb, P. *Word play: What happens when people talk.* New York: Knopf, 1974.

Givens, R. A. The 'plain English' law. *New York State Bar Journal*, 1978, *October*, 479–513.

Goffman, E. *The presentation of self in everyday life.* Garden City, N.Y.: Doubleday, 1959.

Gray, G. W., & Wise, C. M. *The bases of speech* (3rd ed). New York: Harper & Row, 1959.

Guirard, P. *Semiology.* Translated by George Gross. Boston: Routledge and Kegan Paul, 1957.

Hall, E. T. *The silent language.* Garden City, N.Y.: Doubleday, 1959.

Hall, E. T. *The hidden dimension.* Garden City, N.Y.: Doubleday, 1966.

Hartnack, J. *Wittgenstein and modern philosophy.* Translated by Maurice Cranston. Garden City, N.Y.: Doubleday, 1962.

Katz, J. *Semantic theory.* New York: Harper & Row, 1972.

Kendon, A. Review of Mary Ritchie Key. In *Nonverbal communication: A research guide and bibliography. Ars semiotica: International Journal of American Semiotica*, 1976, *2*, 90–92.

Key, M. R. *Nonverbal communication: A research guide and bibliography.* Metuchen, N.J.: Scarecrow Press, 1977.

Korzybski, A. *Science and sanity: An introduction to non-Aristotelian systems and general semantics.* Lancaster, Pa: International Non-Aristotelian Library Publishing Company, 1933.

Korzybski, A. *Science and sanity: An introduction to non-Aristotelian systems and general semantics* (2nd ed.). Lancaster, Pa.: International Non-Aristotelian Library Publishing Company, 1941.

Kübler-Ross, E. *On death and dying.* New York: Macmillan, 1972.

Lieberman, P. *On the origins of language.* New York: Macmillan, 1975.

Minnis, N. (ed.). *Linguistics at large.* New York: Viking, 1971.

Morris, W. (ed.). *The American heritage dictionary of the English language.* Boston: Houghton Mifflin, 1969.

Newman, J. B. General semantics and academic phagocytosis. *Quarterly Journal of Speech,* 1961, *47,* 158–163.

Platt, J. T., & Platt, H. K. *The social significance of speech.* New York: American Elsevier, 1975.

Saussure, F. de. *A course in general linguistics.* W. Baskin, Trans. New York: Philosophical Library, 1959.

Schaff, A. *Introduction to semantics.* New York: Pergamon Press, 1962.

Sheldon, W. H. *The varieties of human physique.* New York: Harper & Row, 1940.

Sheldon, W. H. *The varieties of temperament.* New York: Harper & Row, 1942.

Smith, H. L., Jr., & Ferguson, C. A. *Language and culture.* Washington, D.C.: Department of State Foreign Service Institute, 1951.

Trager, G. L. Paralanguage: A first approximation. *Studies in Language,* 1958, *13,* Nos. 1–2.

Watzlawick, P. *How real is real?: Confusion, disinformation, communication.* New York: Vintage, 1977.

Winter, R. *The smell book.* Philadelphia: Lippincott, 1976.

Wittgenstein, L. *Tractatus logico-philosophicus.* Translated by D. F. Pears and B. F. McGuiness. London: Routledge and Kegan Paul, 1976.

21

Why Do Children Talk?

Mardel Ogilvie

Professor Emerita, Department of Speech and Theater, Herbert H. Lehman College of the City University of New York

Obviously school-age children do not acquire speech and language in a vacuum but rather in the context of the total communicative situation. This situation involves: (1) the children, their friends, classmates, parents, and teachers; (2) the inherent functions of the communicative act including intent; (3) the interests and knowledge of the participants; and (4) the setting. This paper is concerned primarily with the various functions of the communicative act.

Most authorities today in the field of speech communication, whether they are writing about instructing adults or about the language of children, emphasize communication as a shared experience with an intent to influence listeners. For example, Wilson and Arnold (1976), in a text designed for adults note:

> A peculiarity of speech as communication is that it takes place in a particular shared time, at a particular location, and for particular reasons that bring speaker and listeners together. To be successful, then, you as speaker are forced to direct your speaking toward the time and place and reasons for being together, as well as toward whatever other experiences your listeners share with you. (p. 12)

Similarly, Wilkinson (1971), in a book about the language of children, says:

> It is, of course, clear that no language can exist outside a situation—even sentences quoted as examples in a book of this kind are in the situation of

being in a book, and are determined by the points the author wishes to make. But that might be called an artificial situation; nearly all other language occurs in natural situations. And the situation has a considerable effect on the language. (p. 36)

Wilkinson (1971, pp. 105–108) goes on to note the importance of verbal context and then talks about the uses of language. He first explains conative uses—the child makes his own needs known, influences others, gains his own ends, makes it clear that he matters. Wilkinson then continues explaining other uses: affective ones (as in a relationship between mother and child) and cognitive ones such as requests for information, explanations, and definitions and analogies.

Again, similarly, Wilson and Arnold (1976, pp. 153–154) indicate that some of the purposes of adult speakers are to inform, to persuade, to induce inquiry, and to entertain. Children, as we are aware, speak for these same reasons. Johnny in third grade tells about a deep-sea fishing expedition with his father; later in the day he persuades his classmates to vote for him as class president; still later, he asks the art supervisor to tell him how to achieve perspective in his painting. And finally, at the end of the day, he entertains some friends with a very tall tale of his catching a whale while he was deep-sea fishing.

Johnny happens to use speech and language effectively in many communicative situations. Why? Perhaps partly because he has the kind of parents who promote using language for a variety of purposes; partly because he has already accumulated a good-sized stock of knowledge and experiences; partly because he lives in a neighborhood with children who share many interests. He seems to find himself in situations which invite talk. The particular shared times, the reasons for the communicators to be together, his cognitive background, the particular settings have all influenced his successful use of language. Here the emphasis is on using language purposively; let us examine some of the uses of language.

Halliday (1977, pp. 16–18) studied the language of one particular child from 9 to 18 months. He emphasizes that language evolves in the service of certain particular human needs. He constructed the various strands that make up a pattern of thinking about language in functional terms. Taking into consideration his observations of the child, theoretical considerations about linguistic functions—both those essentially linguistic and those extralinguistic in nature—and lastly sociological theories embodying some concepts of cultural transmission and processes of socialization, he evolved the following set of functions:

1. Instrumental (I want)—language used to satisfy the child's needs or desires, e.g., *I want to go with Daddy.*

2. Regulatory (do as I tell you)—language used to control another person's behavior, feelings, or attitudes, e.g., *Find my doll. Don't be mad at me. Let's go.* The child hears this use of language frequently.

3. Interactional (me and you)—language used in the give and take of social dialogue, particularly with his mother and other important individuals. Includes such items as teasing, conversational greetings, and invitations, e.g., *Hi! How are you?*

4. Personal (here I come)—language used to express the child's own individuality, pride in himself or herself, feelings, e.g., *I did the best drawing.* (Includes expressions of interest, pleasure, disgust, etc.).

5. Heuristic (tell me why)—language used for finding knowledge, e.g., *Tell me what makes the light go on.*

6. Imaginative (let's pretend)—language used in make-believe play, e.g., *I'm a princess* or *I'm the big, bad wolf.* Here, the child creates an environment of his own.

7. Representative or informative (*I've got something to tell you.*)—language used to give information to describe a deed or object; in other words, to represent experiences to others, e.g., *Let me tell you about my trip to the Fair.* (Halliday, 1977, pp. 18–32, 37)

In the above instances, examples have been added to clarify the use of the functions. Halliday believes that these functions develop in approximately the order listed—and in any case, the "informative" is significantly last.

Smith (1977) adds three categories to Halliday's seven. They are:

1. Divertive—language used for enjoyment such as puns, jokes, and riddles

2. Authoritative, contractual—language used to indicate how life must be (stated laws, agreements, contracts)

3. Perpetuating—language used to indicate how life was in the past (records, histories, diaries, notes)

Pinnell (1975) did an on-site investigation examining the language purposes of 12 children in three informal classrooms by recording their language through concealed microphones. Using Halliday's categories, he found that first-graders for almost half of the time used language for interactional function and most of the rest of the time for the regulatory and informative.

Different authorities list different functions of language in communication in children. Allen and Wood (1978) note five: controlling, sharing feelings, informing, ritualizing, and imagining. Winkeljohann (1981) indicates that language serves the following meaningful uses: to convince, to direct, to request, to tell, and to explain; schools, however, stress the informational purpose. Searle (1969) incorporates under illocutionary acts commanding, warning, promising, questioning, stating, imploring, and requesting; and under perlocutionary acts (performed when speech acts succeed in modifying the listener's behavior and beliefs), persuading and alarming.

Klein (1979) includes the following purposes:

1. To inform—to provide information to another as giving reports or directions
2. To move to action—to control, manipulate, or persuade (including commands, orders, pleading, justifying, bargaining)
3. To inquire—to seek information
4. To enjoy—to tell stories, to speculate, to theorize (Klein emphasizes that these activities consist of serious contemplation of issues as well as communication to entertain)
5. To conjoin—to maintain social relationships

The Bullock report (1975) suggests that the teacher in the intervention of language difficulties recognize the following uses of language:

> Reporting on present and recalled experiences.
> Projecting into the future; anticipating and predicting.
> Collaborating toward agreed ends.
> Projecting and comparing possible alternatives.
> Perceiving causal and dependent relationships.
> Giving explanations of how and why things happen.
> Expressing and recognizing tentativeness.
> Dealing with problems in the imagination and seeing possible solutions.
> Creating experiences through the use of imagination.
> Justifying behaviour.
> Reflecting on feelings, their own and other people's. (p. 67)

To summarize, charting these purposes, we find that generally they fall into six categories:

1. Fulfilling one's own needs. Such terms as *regulatory, controlling, commanding, warning, promising, alarming,* and *justifying one's own behavior* can be included here.
2. Informing and asking others for information. The term *inform* occurs most frequently here. Other terms include *heuristic, explaining, inquiring, understanding something, reporting on experiences, giving explanations of why things happen.*

3. Persuading. Terms here vary more widely. They include *persuade, convince, implore, move to action, believe something firmly,* and *accept advice.*
4. Entertaining oneself and others and/or imaginative use of language. *Imaginative, enjoying* are among terms often used. Others include *dealing with problems of imagination, creating experiences through imagination.*
5. Impressing. This purpose was noted infrequently; one of the terms used is *ritualizing.*
6. Talking in social situations. This purpose occurred frequently and included terms such as *interacting, sharing feelings, telling, conjoining,* and *reflecting on feelings.*

Obviously none of these categories is discrete. A second-grader, trying to persuade his classmates to make his dog Fido their mascot, told tales of the dog's daring and high intelligence. At the same time, having read how dogs serve as mascots for fire companies, he showed how Fido could play such a role. Finally, he ended his spiel with a demonstration: He became Fido—demonstrating his dog's cleverness and charm. Although his principal purpose was to persuade, three functions of language—informing, persuading and entertaining—were intertwined in this bit of school work.

Both adults and children must make the intent of their message clear so that the purpose can be achieved. This aspect, like phonological, semantic, morphological, and syntactical aspects of language, becomes more sophisticated as children mature.

Leonard, Wilcox, Fulmer, and Davis (1978) point to this growth. They examined four-, five-, and six-year-old children's understanding of indirect requests using 40 videotaped interactions between adults. The children were asked to judge the appropriateness of a listener's response to three indirect requests seen on the screen: (1) one involving an affirmative syntactic structure, as *Can you shut the door?* (2) one involving a negative element, as *Can't you answer the phone?* and (3) one involving a change in the predicate suggesting a change in behavior, as *Must you play the piano?* Even the youngest child showed an understanding of the first two indirect requests but only the six-year-olds showed an understanding of a request for a change in behavior as indicated in the predicate. Children base their reactions to a message on their knowledge of its intent; they often understand the conveyed meanings of even indirect requests.

Many factors concerning contexts are still to be examined. Parents undoubtedly supply quite different contexts for understanding the import of a message. For instance, a parent may almost never use any of the indirect requests. Rather, the father or mother will say to the child, "Please shut the door." They purposively and consistently use the direct approach. If the child asks "Why," the parent may well respond with "Because I said so," or just "Because." A child with different parents with quite different approaches seems to understand equally well. Their request may take the form, "It would be helpful if you closed the door" or, even looking at the door and remarking, "It's cold in here." Either form may well bring the desired response. But we know little about the effects of differing expressions of intent in communication with parents.

Mothers seem to vary the message even according to the sex of the child. Cherry and Lewis (1976) studied the communication of 12 upper-middle-class mothers playing with their two-year-olds. They found that the mothers of the girls talked more, asked more questions, repeated their children's utterances more often, and used longer utterances than the mothers of the boys. The mothers of the male children used more directives than the mothers of the female children.

Obviously, both parents and children use functions of language as the child is reared. Moerk (1977, p. 147) notes:

> Nearly from the beginning of his life, the infant has been actively involved in manipulating objects and persons in his environment. He himself has been manipulated in caregiving procedures and has observed other persons manipulating or moving in space and performing acts in temporal sequence.

When the purpose of the message is not clear, the listener indicates in some way that he or she does not understand, whereupon the speaker attempts to clarify the meaning. Children clarify their meaning when adults or other children do not respond appropriately to their messages. Gallagher (1977) found that normal children around the age of two, regardless of language study, revise the linguistic form of the message when they perceive that the listener does not understand. In this study, the experimenter, pretending that she did not understand what the child had said, would ask, "What?" The revisions followed this pattern: (1) no response, (2) repetition (the child repeats what was said before the "What?"), (3) revision. The revisions fell into four categories: (a) phonetic change—changing the pronunciation of a word, (b) constituent elaboration—adding a morpheme or a word, (c) constituent reduction—omitting a word or morpheme, and (d) constituent substitution—replacing one word with another. An overwhelming majority of the responses were revisions.

A message is more than the particular words involved or how they are strung together. A word, sentence, and string of sentences may communicate quite different messages according to their different contexts. The words involved or the sounds within the words, or the stringing of them together are not the basic units of communication. Rather, the basic unit is the speech act with its intent. When the message is not clear, we revise the message.

Why, in honor of Arthur Bronstein, have we talked about the functions of language—about getting the intent of the message across to listeners? Admittedly, Arthur Bronstein has always been a superb communicator—whether his purpose was to conjoin or socialize (he and Elsa are social creatures), to inform (always done clearly, succinctly, and graciously), to enquire (he has a natural, inborn intellectual curiosity), to provoke laughter (with one of his totally crazy limericks), or to persuade (more about this in the next paragraph)—or whether his purpose was ritualistic, as when he recently addressed a group assembled for a funeral. Incidentally, a remark overheard after this address was, "A hard act to follow." But no, the intent of this paper is not to extol Arthur Bronstein's communicative abilities. Admittedly, the purpose is devious.

Arthur Bronstein—we need to know how parents, teachers, and children use language for the variety of functions, how, when, and in what circumstances parents, teachers, and children are successful in getting the intent of their messages across to their listeners, and how they react when they are unsuccessful in this endeavor. We need data on the responses of children to messages with different purposes, in different settings, and in different contexts. In our libraries, we find a considerable number of research reports on phonology, morphology, syntax, vocabulary—but a paucity of reports of research on pragmatic aspects of language. Arthur Bronstein, persuasive man that you are, can you entice one graduate student, or two, or maybe even three, into this area of research?

REFERENCES

Allen, R. R., & Wood, B. S. Beyond reading and writing to communication competence. *Communication Education*, 1978, 27, 286–292.

Bullock, A. *A language for life*. Report of the Committee of Inquiry appointed by the Secretary of State for Education and Science. London: Her Majesty's Stationery Office, 1975.

Cherry, L., & Lewis, M. Mothers and two-year-olds: A study of sex differentiated aspects of verbal interaction. *Developmental Psychology*, 1976, 12, 278–282.

Gallagher, T. M. Revision behaviors in the speech of normal children developing language. *Journal of Speech and Hearing Research*, 1977, 21, 303–313.

Halliday, M. A. K. *Learning how to mean: Explorations in the development of language.* New York: Elsevier, 1977.

Klein, M. L. Designing a talk environment for the classroom. *Language Arts,* 1979, *56,* 647–654.

Leonard, L. B., Wilcox, M. J., Fulmer, K. C. & Davis, G. A., Understanding indirect requests: An investigation of children's comprehension of pragmatic meanings. *Journal of Speech and Hearing Research,* 1978, *21,* 528–537.

Moerk, E. L. *Pragmatic and semantic aspects of early language development.* Baltimore: University Park Press, 1977.

Pinnell, G. S. *Language functions of first grade students observed in informal classroom environments.* Unpublished Ph.D. Dissertation, Ohio State University, 1975.

Searle, J. R. *Speech acts: An essay in the philosophy of language.* London: Cambridge University Press, 1969.

Smith, F. The uses of language. *Language Arts,* 1977, *54,* 638–644.

Wilkinson, A. *The foundations of language: Talking and reading in young children.* London: Oxford University Press, 1971.

Wilson, J. F., & Arnold, C. C. *Dimensions of public communication.* Boston: Allyn & Bacon, 1976.

Winkeljohann, R. How can teachers promote language use? *Language Arts,* 1981, *58,* 605–606.

22

The Role of Formant Transitions in the Perception of Stress in Disyllables

Lawrence J. Raphael

Department of Speech and Theater, Herbert H. Lehman College of the City University of New York, and Haskins Laboratories

Michael F. Dorman

Department of Speech, Arizona State University

In a series of experiments reported elsewhere (Raphael & Dorman, 1981; Raphael, Dorman, & Liberman, 1980), we have found that the syllable-initial formant transitions of synthetic CVC syllables contribute by their duration to the perception of vowel duration. Our experimental paradigm called for subjects to classify syllable–final consonants in CVC and VC syllables as English /t/ or /d/. The sole independent variable in the experiments was the duration of the steady-state portion of the vowel. The /t/-/d/ phoneme boundaries, plotted as a function of vocalic (formant) duration, were virtually coincident in both CVC and VC syllables, indicating that the initial formant transitions were accommodated into the listeners' perceptions of duration to about the same extent as the steady-state formants.

Our results were in general agreement with those of Mermelstein, Liberman, and Fowler (1977) and Verbrugge and Isenberg (1978), who performed analogous experiments using vowel identification and duration-matching tasks. Our results differed from theirs, however, in that

249

they found that the formant transitions and steady-state formants did not contribute equally to the perception of duration. Mermelstein *et al.* (1977) reported the transitions to contribute only 50% of their duration; Verbrugge and Isenberg (1978) reported that the transitions contributed over 100% of their duration.

The experiment reported here was designed for three purposes. The first was to test the basic conclusions of our previous experiments, using a perceptual task other than the identification of final-consonant cognates. The task selected was the judgment of stress location in disyllables. The second purpose was to discover whether formant transitions in other than syllable-initial position contribute to the perception of vowel duration. The use of the stress location task permitted us to evaluate the effects of formant transitions in syllable-final, syllable-initial, and intervocalic positions. The final purpose was to seek further information about relative contributions to the perception of duration of formant transitions and steady-state formants.

METHODS AND PROCEDURES

The disyllabic stimuli of this experiment were created by conjoining pairs of synthetic syllables (Figure 1). These syllables, generated by the Haskins Laboratories parallel resonance synthesizer, contained steady-state formants appropriate to the vowel /a/. One stimulus type (V) consisted of the steady-state segment alone. The other three stimulus types contained initial, final, or initial and final, 50-ms formant transitions appropriate to cue the perception of the stop-consonant /b/. They were a CV: /ba/, a VC /ab/, and a CVC /bab/. Each syllable was then sequentially paired with itself and with the other syllable types, with the exception that no disyllable was created without at least one set of formant transitions at the juncture of the syllables.[1]

A 70-ms silent gap, simulating stop closure, separated the two syllables of each stimulus. In one series of tests the duration of the steady-state portion of the second syllable of each stimulus type was varied from 100 to 200 or 220 ms in 10-ms steps. The first syllable was fixed at a duration of 150 ms. In a second series of tests the duration of the steady-state portion of the first syllable was made to vary in 10-ms steps from 110 to 200 or 220 ms, while the duration of the second syllable was held constant at 150 ms. Since the intensity and fundamental frequency were held constant across both syllables, the only effective variable, for the purpose of cueing stress, was steady-state formant/syllable duration.

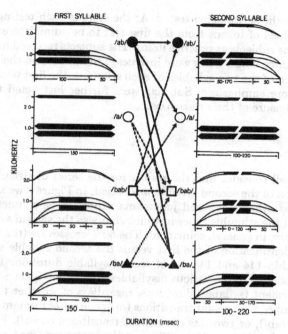

FIGURE 1. Formant frequency values and combination patterns for the synthetic disyllables used as experimental stimuli in which the second syllable varied in duration. The same values and combination patterns were used for the stimulus series in which the first syllable varied in duration.

Four tokens of each disyllabic stimulus type at each duration were recorded in random order on tape for presentation to the subjects, 12 undergraduate students at Herbert H. Lehman College of the City University of New York. Each stimulus type was presented in a separate test. None of the students had previously participated in listening tests of this type, and none had a history of hearing impairment or speech disorders. They heard the test tapes over an AR–4X loudspeaker in a large sound-attenuated room. The tests were presented in two sessions, a week apart. One group of six subjects heard the first series of stimuli (second syllable-duration varying) during the first session and the second stimulus series (first syllable-duration varying) the following week. The other group of six subjects heard the stimulus series in the reverse order. During each session there was a five-minute rest period between the tests of each stimulus type within the series being tested. At the start of each testing session a randomized set of tokens from the first test to be administered was presented to the subjects as practice items. The subjects received no feedback for the practice items. They were instructed to indicate on their response sheets which of the two syllables of each stimulus, the first or the second, was the more emphasized. Subjects were further instructed to guess if they were unsure of their responses.

RESULTS

We shall consider first the results for the series of stimuli in which the duration of the second syllable was varied. In Figure 2 we see the percent of first-syllable stressed judgments expressed as a function of the duration of the stimulus types in which /ba/ was the second syllable. The functions are, in general, coincident. The 50% crossovers (the boundaries between identification of the first versus the second syllable as stressed) all fall within 144 and 148 ms of second-syllable duration. That is, listeners responded to the various disyllables almost purely on a basis of the relative formant (syllable) durations, regardless of whether the first syllable contained no formant transitions (/a/), one set of formant transitions (/ba/ and /ab/), or two sets of formant transitions (/bab/). Further, the placement of the formant transitions at the beginning (/ba/ + /ba/) or at the end (/ab/ + /ba/) of the first syllable did not differentially affect the listeners' responses in any important way.

We find comparable results for the other stimulus types of the first

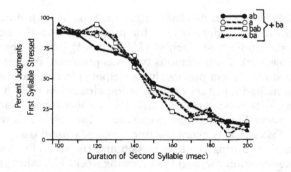

FIGURE 2. Percentage of trochee judgments for disyllabic stimuli as a function of variation in the duration of the second syllable: /ba/.

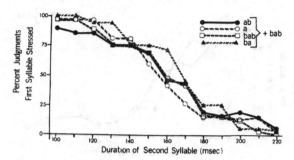

FIGURE 3. Percentage of trochee judgments for disyllabic stimuli as a function of variation in the duration of the second syllable: /ab/ and /a/.

series. In Figure 3 we see that the stress boundaries for the stimuli ending with the syllable /bab/ lie between 154 and 168 ms of second-syllable duration. Once again the functions largely coincide. Figure 4 shows the results for the stimuli ending in /ab/ and in /a/. The stress boundaries range between 158 and 162 ms of second-syllable duration.

The results for the second set of stimuli, in which the duration of the first syllable was varied, were much the same as those for the first stimulus series. Figure 5 shows a representative sample of the data for the stimuli in which the second syllable was /ba/. We see that the stress boundaries for the /ab/ + /ba/ and the /a/ + /ba/ stimuli lie at 143 and 150 ms of first-syllable duration and that the functions, once again, are generally coincident.

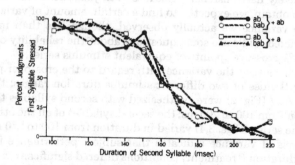

FIGURE 4. Percentage of trochee judgments for disyllabic stimuli as a function of variation in the duration of the second syllable: /bab/.

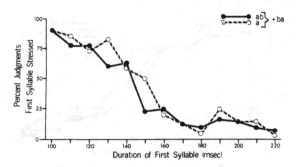

FIGURE 5. Percentage of trochee judgments for disyllabic stimuli as a function of variation in the duration of the first syllable: /ba/.

DISCUSSION

The results of the present experiment confirm the findings of our earlier studies: formant transitions contribute about equally with steady-state formants to the perception of syllable duration. Further, the location of the transitions within the syllable does not affect the extent to which listeners use them when estimating syllable duration: syllables of equal duration affect listeners' judgments equivalently, regardless of whether they comprise CV, VC, CVC, or steady-state formant patterns.

Here we must append a note about the nature of listeners' judgments, which, although it does not affect our conclusions, may serve as a cautionary note to experimenters who wish to elicit reliable judgments of stress from relatively naive listeners. Because such judgments are known to be difficult to obtain, we expected to find a certain amount of variance in our data. The variance we actually observed, though less than might have been predicted, did raise some questions about the reliability of the data for the 50% crossover points of equivalent stimulus series.

In fact, most of the variance with regard to the crossover points was caused by the use of two different stimulus duration ranges: the stimuli ending in /ba/ (Fig. 2) were synthesized with second syllables that varied in duration from 100 to 200 ms; the second syllables of all the other stimuli of the same series (Figs. 3,4) varied in duration from 100 to 220 ms. A one-way analysis of variance indicates that subject performance in the two different duration (treatment) conditions differed significantly: $F(9,99) =$ 3.22, $p. < .005$. Post-hoc tests according to Scheffé reveal that all significant differences are between treatments tested on the 100–200 ms continuum and those tested on the 100–220 ms continuum. Thus the differences between treatments were conditioned by the range of the stimulus continuum rather than by the stimulus manipulation. It appears that some portion of our subjects was relying on perceptions of overall stimulus duration (rather than on relative syllable duration) when responding. (A series of succeeding pilot experiments revealed that some subjects perceptually halved the continuum into two categories when it was extended to 300 ms.) Since all syllable types were judged equivalently, regardless of their durations, we can allow our conclusions concerning the processing of formant transitions to stand as we have stated them above. We cannot, however, be certain to what extent the data supporting those conclusions reflect perceptions of stress as opposed to perceptions of absolute duration.

NOTES

1. The disyllables synthesized as experimental stimuli were thus:

$$
/ab/ + \left\{ \begin{array}{l} /ab/ \\ /a/ \\ /bab/ \\ /ba/ \end{array} \right. \qquad /bab/ + \left\{ \begin{array}{l} /ab/ \\ /a/ \\ /bab/ \\ /ba/ \end{array} \right. \qquad \begin{array}{l} /a/ + \left\{ \begin{array}{l} /bab/ \\ /ba/ \end{array} \right. \\[1em] /ba/ + \left\{ \begin{array}{l} /bab/ \\ /ba/ \end{array} \right. \end{array}
$$

Those disyllables excluded from the stimulus set were:

$$
\left. \begin{array}{l} /ba/ \\ /a/ \end{array} \right\} + /a/
$$

$$
\left. \begin{array}{l} /ba/ \\ /a/ \end{array} \right\} + /ab/
$$

REFERENCES

Mermelstein, P., Liberman, A. M., & Fowler, C. Perceptual assessment of vowel duration in consonantal context and its application to vowel identification. *Journal of the Acoustical Society of America*, 1977, *62*, S101 (Abstract).

Raphael, L. J., Dorman, M. F., & Liberman, A. M. On defining the vowel duration that cues voicing in final position. *Language and Speech*, 1980, *23*, 297–307.

Raphael, L. J., & Dorman, M. F. The contribution of CV transition duration to the perception of final consonant voicing in natural speech. *Journal of the Acoustical Society of America*, 1980, *67*, S51 (Abstract).

Raphael, L. J., & Dorman, M. F. The contribution of extended CV transitions and aperiodic formant structure to the perception of duration. *Journal of the Acoustical Society of America*, 1981, *70*, S34 (Abstract).

Verbrugge, R. R., & Isenberg, D. Syllable timing and vowel perception. *Journal of the Acoustical Society of America*, 1978, *63*, S4 (Abstract).

23

Aspects of Deixis in the Language of Children with Autism and Related Childhood Psychoses

Norma S. Rees

Vice Chancellor for Academic Affairs, University of Wisconsin-Milwaukee

Focus on language as an instrument of communication has recently characterized the study of language and of its development in normal and nonnormal populations. When the subject of concern is how sentences actually function in conversation between two or more speakers, the aspect of language known as *deixis* takes on renewed interest. If spoken sentences are to be fully communicative, they must provide information that allows the listener to relate the content of the utterance to the relevant context; among these contextual factors are person, place, and time. Natural languages universally have mechanisms for indicating these relationships.

In the case of person deixis, the speech act must distinguish among the speaker, the person to whom the utterance is addressed, and other person or persons who may be spoken about but who are neither speaker or listener in the speech event under consideration. In English, of course, these distinctions are provided for largely by the system of first, second, and third person pronouns; an illustration appears in (1).

(1) I left you money for the paper boy. Please give it to him when he comes around.

The underlined lexical items in (1) mark distinctions among the persons of the conversation, the speaker, the hearer, and the person referred to.

In the case of place deixis, the speech act must furnish information about where the speaker and hearer are relative to each other and the situational context. Such linguistic units as *this* and *that*, *here* and *there*, *come* and *go*, *in front of* and *behind*, to name but a few, perform this function in English.

(2) Give me that umbrella and keep this one for yourself.

In (2), the underlined words indicate which of two umbrellas present in the situation the listener is to keep, in this case the one closer to the speaker at the time of utterance.

(3) Take the charcoal with you when you go out.

The use of *take* and *go* in (3) implies that the speaker, the listener, and the charcoal are all in the house at the time of utterance (or at least that the speaker is imagining such a state of affairs). If the speaker were out back of the house setting up the barbecue shouting to the listener inside the house, the speaker would say:

(4) Bring the charcoal with you when you come out.

Place deixis, therefore, encodes information about the relative locations of speaker, listener, objects referred to, and setting.

In the case of time deixis, the speech act must contain information about the time of the action or event referred to relative to the time of the utterance itself. Minimally, the speech act distinguishes between the now and the not-now and in English expands to a system for indicating past, present, and future. Linguistic devices like *now* and *then, yesterday, today,* and *tomorrow,* and the verb tense system all perform this function in English.

(5) The car crashed into the cyclist.

(6) Daddy will be home soon.

Moreover, in (6) *soon* specifies that the future event will take place after only a relatively small segment of time has elapsed.

Linguistic deixis of person, place, and time are all illustrated in (7), an utterance that is plainly uninterpretable without knowledge of the linguistic markers for deixis.

(7) I'll have been there for an hour by the time you arrive.

While the specific details of the deictic system vary considerably from language to language, deixis of person, place, and time are characteristic features of natural languages (Greenberg, 1966). The few simple illustrations of the above summary provide only a superficial overview of the English language's deictic system, which is both highly flexible and very complex. That deixis is an inescapable aspect of language in communication is dramatically reflected in the recent analyses of sign language of the deaf. Specifically, American Sign Language, a naturally occurring and naturally learned language in its own right, utilizes hand movements relative to the speaker's own body to perform deictic functions. In ASL person deixis is marked by pointing toward the signer, the person addressed, and in an outward direction, thus affording the first-, second-, and third-person distinctions; place deixis, by pointing down for *here* and away from the signer, in various movements and orientations, for *there*; and deixis of time, by movement pointing behind the signer's body for past, the space occupied by the body for present, and the space in front of the body for future (Friedman, 1975).

According to Rommetveit (1968), deixis is the basic mechanism for bringing information from the nonlinguistic context of the speech event into the spoken message. The system for relating the utterance to the context of person, place, and time permits what Rommetveit calls the "deictic anchorage" without which sentences cannot function in communication. In a broader sense, all linguistic devices for relating sentences to linguistic and nonlinguistic context are implicated by this concept, and Fillmore (1971) adds two deictic categories to the usual three of person, place, and time:

> An enriched and extended account of deixis for natural languages can be constructed by adding to the traditional list such matters as (iv) *discourse deixis*, by means of which a communication can refer to a portion of the ongoing discourse; and (v) expressions which reflect the social relationship which the speaker regards as holding between the interlocutors, especially as these are codified in systems of honorific or deferential speech.

What Fillmore calls *discourse deixis* appears to be covered by the more current term *cohesion* (Halliday & Hasan, 1976); elsewhere Fillmore (1975) uses the term *role deixis* to stand for the fifth type, the one that reflects social relationships. For the purposes of this paper, only the traditional categories of person, place, and time deixis will be considered. To summarize these categories, the deictic system permits the speaker to express, and the listener to comprehend, who is speaker, who is listener, who is being talked about, where these persons are relative to one another and their surroundings, and when the action spoken about occurs relative to the time of speaking.

A key element in the deictic system for person, place, and time is that the speaker is the central point of reference throughout (Fillmore, 1971, 1975; Clark, 1978). This concept can be captured by noting that common to all three types of deixis is a minimal contrast between the speaker and something else: the speaker versus the listener (in person deixis, illustrated by *me* and *you*); the location of the speaker versus any other location (in place deixis, illustrated by *here* and *there*); and the time of the speaking versus any other time (in time deixis, illustrated by *now* and *then*). While the total system of linguistic deixis is far more elaborate than these examples of binary contrasts reflect, it is nonetheless reasonable to describe the role of the speaker as the pivotal point of reference around which the system operates. The illustrations of deictic devices in American Sign Language, given above, dramatize this feature. In ASL the phonology of spoken language is replaced by a manual-visual system of hand movements relative to the signer's body (Friedman, 1975), and the indicators of person, place, and time visibly revolve around the signer as point of reference. The centrality of the speaker (or signer) to the system itself is an important basis to the speculations that follow.

Thus far it is established that the deictic system is an essential factor in the use of language in communication; that the system is both flexible and complex; that it includes (minimally) the categories of person, place, and time; and that the speaker is at the center of the system. The implications for analysis of the language of children with autism and related childhood psychoses may be taken up next. From a narrow perspective, certain clinical language findings may be aptly described as deictic failures. A relevant example is the well-known tendency of autistic children to reverse the pronouns *I* and *you* (Fay, 1971, 1979). That this difficulty with pronouns is not the only relevant phenomenon was pointed out by Bartolucci and Albers (1974), who noted difficulty in the use of deictic markers for time (specifically, past tense) among autistic children. The conclusion that the language of autistic children may be deficient with respect to mastery of the deictic system is compatible with the clinical experience that the spoken language of these children typically is not communicative. Even more interesting, however, is the possibility that the nature of these childhood disorders itself accounts for the failure to develop a functional deictic system.

To begin with, it is necessary to define the child population under consideration. The literature on infantile autism is filled with disputes about whether it is or is not a discrete syndrome, diagnostically distinguishable from such disorders as childhood schizophrenia, mental retardation, and central nervous system disorder. Without attempting to resolve definitional controversies, here autism will be included among the relevant childhood psychoses, either primary in nature or secondary to other processes such as chronic brain syndrome or sensory deficits (especially blindness). The source of this classification is the work of Menolascino (1970), who concludes that infantile autism is not a distinct diagnostic category but rather a subtype of childhood psychosis. He defines the latter disorder as one "in which there is developmental failure (or dissolution) of the self-concept system that is consonant with the meaningful personal-social responses that ordinarily act to conserve both the identity and integration of the personality" (page 129). These children may, in some instances, fail to develop a functionally complex ego, resulting in the typical clinical description of "sparse reciprocal interpersonal relationship," lack of discrimination between animate and inanimate objects, lack of eye contact, and language deficiencies ranging from nonverbal status to complex but uncommunicative use of language. Fay (1979) also describes the autistic child as one who is handicapped both in the perception of self and the perception of social relationships. Some of these children talk and others do not, but Churchill (1978) asserts that language dysfunction to one degree or another is common to all children with a diagnosis of infantile autism; probably the same can be said with respect to children defined according to Menolascino's broader classification of childhood psychosis.

In normal child development, it appears that the origins of the self–other distinction may be also the origins of deixis. There has been much interest of late in the early interaction between infant and caregiver as the source of these developments. An eloquent introduction is provided by Erik Erikson:

> The self-images cultivated during all the childhood stages thus gradually prepare the sense of identity, beginning with that earliest mutual recognition of and by another face which the ethologists have made us look for in our human beginnings. Their findings, properly transposed into the human condition, may throw new light on the identity-giving power of the eyes and the face which first "recognize" you (give you your first "Ansehen"), and new light also on the infantile origin of the dreaded estrangement, the "loss of face." To be a person, identical with oneself, presupposes a basic trust in one's origins—and the courage to emerge from them. (pages 94–95)

The pragmatic processes of the early period are described further by Bruner (1975b), who emphasizes the significance of mother–infant interaction in the infant's development of reversible participant-roles. Bruner calls attention to the apparently universal interactive play routines that mothers and infants perform with much repetition and mutual delight. For example, the ubiquitous give-and-take game which mothers engage in with infants just old enough to sit up is typical of the format in which each plays a complementary role to that of the other but where roles can be, and are, reversed, as the game proceeds. Bruner stresses the use of early sound-making in interactive routines in promoting shared attention and joint action between infant and caregiver, reporting that infants as young as four months can follow the mother's line of regard to locate an object of mutual interest. These early experiences, then, give to the infant not only a self and an identity but also the distinction between self and other without which there can be no self. The beginnings of the ability to take the other's perceptual standpoint also may be seen in these early interactive, reversible, games. Bruner (1975a) concludes that "a grasp of reciprocal roles in discourse is the essential prerequisite for deixis of person, place, and time." These descriptions are consistent with the account of the developing deictic system wherein the speaker is the reference point with which concepts of person, place, and time may be matched or contrasted.

Knowledge of the developmental aspects of deixis is as yet fragmentary both for normal and disordered children. Even normal children, of course, have to learn the specific linguistic devices their community's language makes available for marking deixis, and they do not acquire them all at once. Clark (1978) proposes that deictic terms might be more difficult for children to learn because their properties of shifting reference (*you* becomes *I* when roles are reversed, but proper names remain stable) and shifting boundaries (*here* and *there* are imprecise and variable, in contrast to explicit location or time). A number of investigators have commented on the developing mastery by preschool children of specific linguistic elements that express deictic concepts. Bates (1976) and Clark (1978) trace the origin of deictic words like *this* and *that* to the child's use of gesture (pointing) to secure the adult's attention to an object of joint interest. Maratsos (1974) states that the child's early use of pronoun forms has a deictic, or "pointing out" character. According to Clark (1978) and Clark and Sengul (1978), children's earliest use of deictic terms is noncontrastive, that is, the child says *da (that)* to indicate any object, whether close to the speaker or far away, and over time learns to use *this* and *that* to distinguish between objects in terms of relative distance from the speaker. In a comprehension task, she reports full contrastive use of *this* and *that*, *here* and *there* beginning at around five years. DeVilliers and deVilliers, however, report that three- and four-year-olds used these words correctly in a comprehension task.

Acquisition of the verbs *come* and *go*, *bring* and *take* was studied by
Clark and Garnica (1975) in comprehension tasks carried out by children
between 5½ and 9½ years; these authors concluded that children acquire
the full adult meanings in stages. Children's nonlinguistic strategies
account, predictably, for comprehension errors prior to full mastery of the
word meanings. Younger children, around four and five years, do not fully
distinguish the meanings of *come* and *go* in comprehension, according to
Clark (1978), but Macrae (1976) found that two-year-olds do contrast
these words in spontaneous utterances.

Oddly, the models for the deictic system children are exposed to in
ordinary interaction with caregivers and other adults deviate from the
adult system in a number of predictable ways, raising questions about how
children learn the adult form at all. The deviations are especially notable
in the case of person deixis in "baby talk," the special way that adults talk
to human infants (and sometimes to other addressees who remind them
of human infants—pets, lovers, hospitalized adults, even plants). The
major purpose of using baby talk is believed to be to facilitate communication with a listener who has difficulty understanding ordinary adult
speech, although the modifications of baby talk may be adapted to other
circumstances like that of lovers (Brown, 1977; Ferguson, 1977). While
baby talk's consistent features involve aspects of prosody, phonology, syntax, and the lexicon, the pertinent aspects here are the uses of deictic pronouns for person. In her groundbreaking analysis, Wills (1977) shows how
the pronoun system of baby talk differs from that of the adult language,
a few examples of which follow.

1. The speaker refers to herself/himself in the third person:
    (8) Mommy's busy now, darling.

2. The speaker refers to herself/himself in plural:
    (9) We'll just put your socks on.

3. The speaker refers to the listener in the third person:
    (10) Does baby want a cookie?

4. The speaker refers to the listener in first person plural:
    (11) We're eating up all our breakfast.

Wills furnishes 11 categories of baby talk pronoun usage of which the
above four are perhaps the most common. If the major purpose of baby
talk modifications is to promote comprehension with a less than fully
accomplished language user, it might be that referring to the speaker or
listener by name or kin relationship or common noun can be explained on

the basis that it is easier to understand than the I/you contrast requiring, as that does, a shift in perspective. It might be argued that, for younger listeners, speakers adopt the baby talk version of person deixis as a means of insuring comprehension as well as a device for teaching role and kinship relationships. Ferguson (1977) and Wills (1977) suggest this interpretation, pointing out that adults gradually abandon the baby talk pronoun usage as children become more competent linguistically. That argument fails to explain the modifications wherein *I* becomes *we* or *you* becomes *we*, and Wills's explanation is now different: she suggests that *we* in place of *I* may express the meaning of involving the child in an action that is in reality performed by the parent alone. *We* in place of *you*, however, is explained as a playful or cajoling alternative for what might otherwise have a punitive quality. Apparently, then, ways in which the baby talk pronoun system deviates from the adult system are not a simple matter of ease of communication and are not anything like the devices for teaching the child linguistic or social conventions. While the work on baby talk uncovered interesting developmental aspects of deixis, the analyses reported do not explain how children acquire the adult version of the system.

To return to the specific subject at hand: the way in which autism and related childhood psychoses interfere with the development of the system of linguistic deixis may not, after all, be primarily a matter of atypical patterns of development of specific deictic terms. Of primary interest to the application of deixis to the language development of autistic and psychotic children may, in fact, be the prelinguistic period. Fundamentally, the very basis for development of deixis may be disrupted in the early prelinguistic stages of development. Children whose early lives fail to provide the early experiences of joint attention and interactive play with an adult caregiver like those described by Bruner, or children who are unable to profit from these kinds of experiences because of their own inherent malfunctions, may be expected to manifest disorders of self-concept; among such children, the ones who do learn to talk appear to be prone to a developmental difficulty in mastering linguistic deixis.

It is possible that the impairments of deictic development associated with autism and related childhood psychoses are fundamentally failures of the basis for establishing person deixis, and only secondarily failures of place and time deixis. That is, assuming that the self as differentiated from the other is undifferentiated or weakly differentiated in such children, the speaker–self around which person deixis develops and from which deictic contrasts eventually emerge is unavailable or too unstable to serve as central reference point. In this connection, it is noteworthy that Clark (1978) considers the *I–you* distinction the easiest (i.e., simplest) of all deictic contrasts the adult system provides. From that cognitive standpoint, then, if deictic development breaks down at the *I–you* distinction, the remainder of the system would probably not develop effectively. We may make a somewhat stronger claim, that the child in whom development of the self is uncertain may, more or less in order of degree of language impairment, (1) not learn to talk at all; (2) fail to learn reference; (3) talk very little; (4) talk solely or primarily in overlearned, ritualistic, or echolalic responses; (5) be unable to participate in reciprocal role situations; (6) make many inappropriate utterances; (7) talk intelligibly but with unclear or no apparent communicative purpose; or (8) fail to develop complete control of the adult system of linguistic deixis. Some examples may be found in Fay (1971, 1979), Simon (1975), Miller (1978), and Churchill (1978).

These considerations offer, moreover, a way of describing some perplexing instances of the use of language by children with autism and related disorders of childhood psychosis. Of particular interest are those utterances taken from dialogue with, for example, a teacher, where the child participates in an intelligible, perhaps semantically related (to the context) fashion without, however, encoding his or her remarks in expected form. While the meanings of these utterances may be comprehensible to the listener, the form of encoding makes it seem as if the child were speaking from the standpoint of someone else talking to or about him or her. These examples from Debbie, a blind ten-year-old who is psychotic, are illustrative:

(12) Teacher: Push it in.
Debbie: Push. O.K.
Teacher: What did you say?
Debbie: She said push.

(13) Teacher: Do you have any water in your pocket?
Debbie: I have any water in her pocket.
Debbie: (bending down to scratch an itch) Who's got something on her leg?

(14) Teacher: Say it again Deb.
Debbie: She didn't say it. Watch Debbie. She made a noise.

Debbie has learned the language—her utterances are generally grammatical and semantically related to the linguistic and situational context—but they give the impression of coming from a speaker with no communication role of her own. The deictic devices for expressing contrast between speaker and other are missing or inconsistent. Children like this often utilize overprogrammed, overconditioned segments of spoken language to express meaning, as in the case of Thomas, an eight-year-old who like Debbie is blind and psychotic: When the lunch bell rings, Thomas comments:

(15) Time for lunch now!

but, in contrast, never says "I'm hungry" upon hearing the lunch bell or, for that matter, at any time.

Even more startling are the instances when the child takes both parts in a conversational interchange, as in the case of Jackie, a 13-year-old, participating in a group session where the teacher has just said, "Come on, everybody, let's sing a song." Jackie's response:

(16) I don't have to sing if I don't want to. Yes, you do.

In summary, linguistic deixis is essential to the use of language in communication. Like self-concept itself, the deictic system appears to originate in the early interactive, reciprocal experiences between infant and other. It may be said that the deictic system is the basic linguistic expression of the self–other distinction. Although breakdowns in the use of linguistic deixis contribute heavily to the uncommunicative quality of the spoken language of children with autism and related disorders of childhood psychosis, it is also the case that failure to develop a stable self-concept will be associated with varying degrees of developmental disorders of language.

REFERENCES

Bartolucci, G., & Albers, R. Deictic categories in the language of autistic children. *Journal of Autism and Childhood Schizophrenia*, 1974, *4*, 131–141.
Bates, E. *Language and context: The acquisition of pragmatics*. New York: Academic Press, 1976.
Brown, R. Introduction. In C. E. Snow & C. A. Ferguson (Eds.), *Talking to children: Language input and acquisition*. Cambridge, England: Cambridge University Press, 1977.
Bruner, J. S. From communication to language: A psychological perspective. *Cognition*, 1975, *3*, 255–287.(a)
Bruner, J. S. The ontogenesis of speech acts. *Journal of Child Language*, 1975, *2*, 1–19.(b)
Churchill, D. W. *Language of autistic children*. New York: Wiley, 1978.
Clark, E. V. From gesture to word: On the natural history of deixis in language acquisition. In J. S. Bruner & A. Garton (Eds.), *Human growth and development*. Oxford: Oxford University Press, 1978.

Clark, E. V., & Garnica, O. K. Is he coming or going? On the acquisition of deictic verbs. *Journal of Verbal Learning and Verbal Behavior*, 1975, *13*, 559–572.

Clark, E. V., & Sengul, C. J. Strategies in the acquisition of deixis. *Journal of Child Language*, 1978, *5*, 457–475.

deVilliers, P. A., & deVilliers, J. G. On this, that, and the other: Nonegocentrism in very young children. *Journal of Experimental Child Psychology*, 1974, *18*, 438–447.

Erikson, E. H. *Insight and responsibility*. New York: Norton, 1964.

Fay, W. H. On normal and autistic pronouns. *Journal of Speech and Hearing Disorders*, 1971, *36*, 242–249.

Fay, W. H. Personal pronouns and the autistic child. *Journal of Autism and Developmental Disorders*, 1979, *9*, 247–260.

Ferguson, C. A. Baby talk as simplified register. In C. E. Snow & C. A. Ferguson (Eds.), *Talking to children: Language input and acquisition*. Cambridge, England: Cambridge University Press, 1977.

Fillmore, C. J. *Toward a theory of deixis*. (PC CLLU papers, 3.4.) Honolulu: University of Hawaii, 1971.

Fillmore, C. J. *Santa Cruz lectures on deixis, 1971*. Bloomington, Indiana: Indiana University Linguistics Club, 1975.

Friedman, L. A. Space, time, and person reference in American sign language. *Language*, 1975, *51*, 940–961.

Greenberg, J. H. (Ed.). *Universals of language*. Cambridge, Mass.: M.I.T. Press, 1966.

Halliday, M. A. K., & Hasan, R. *Cohesion in English*. London: Longmans, 1976.

Macrae, A. J. Movement and location in the acquisition of deictic verbs. *Journal of Child Language*, 1976, *3*, 191–204.

Maratsos, M. P. Preschool children's use of definite and indefinite articles. *Child Development*, 1974, *45*, 446–455.

Menolascino, F. J. Infantile autism: Descriptive and diagnostic relationships to mental retardation. In F. J. Menolascino (Ed.), *Psychiatric approaches to mental retardation*. New York: Basic Books, 1970.

Miller, L. Pragmatics: *An assessment/intervention model used with an autistic child*. Paper presented at the annual meeting of the American Speech-Language-Hearing Association, 1978.

Rommetveit, R. *Words, meanings and messages*. New York: Academic Press, 1968.

Simon, N. Echolalic speech in childhood autism. *Archives of General Psychiatry*, 1975, *32*, 1439–1446.

Wills, D. D. Participant deixis in English and baby talk. In C. E. Snow & C. A. Ferguson (Eds.), *Talking to children: Language input and acquisition*. Cambridge, England: Cambridge University Press, 1977.

24

Jonathan Boucher's Farewell Sermon

John F. Wilson

Department of Speech and Theater, Herbert H. Lehman College of the City University of New York

> Sermonizing has become an indispensable and valuable part of our pattern of worship. It has been so from times that fostered our ancient heritage. It may change in style or mood or form. Speeches that are carefully conceived, developed and effectively presented are hardly outmoded, certainly not obsolete. Excising them will remove a central, most valuable function of the preacher. He is a leader of a social group and he leads and teaches in acts and words. He cannot comfortably break this mold. For it is not tradition that demands the worthwhile sermons. It is people.
>
> —Arthur J. Bronstein (1964)[1]

On Sunday, July 24, 1775, the Reverend Jonathan Boucher entered his church, Queen Anne's Anglican Church in St. George's county, Maryland, mounted his pulpit, and placed a pair of pistols on the cushion. Then, in what has become known as his "Farewell Sermon," he shouted down his detractors and preached a diatribe against participation in rebellion.

The previous Thursday, July 21, had been a day set aside for worship and prayer by resolution of the Second Continental Congress. Boucher appeared at his church for services on that day knowing that his presence would be deeply resented by the Whig leadership of the parish. He was stunned to find "not less than 200 armed men" (Boucher, 1925, p. 121) crowded into the church intent on preventing him from preaching. He sent word to the crowd's leader that the pulpit was his by right and that he intended to occupy it at the risk of his life.

For the preceding six months, Walter Harrison, Boucher's curate, had conducted the services at Queen Anne's. The pastor had withdrawn to serve in a less hostile parish where his wife's uncle was rector. But even there his advocacy of "peaceableness" (p. 113) gave offense and he had begun the practice of placing a pair of loaded pistols on the cushioned seat of the pulpit (p. 113). He had chosen to return to Queen Anne's on the official day of prayer and fasting in behalf of the colonial cause after learning that Harrison was "promoting factious" activities (p. 119) in support of colonial resistance. He knew that a sermon preached on that day would receive the widest possible hearing. His friend Governor William Eden of Maryland had urged him to appear (p. 119).

On Thursday, Boucher went to the church and told Harrison that he intended to preach that day and urged him "at his peril, not to attempt to dispossess me" (p. 121). He took a seat nearer the pulpit than Harrison. In his own words, "At the proper time, with my sermon in one hand and a loaded pistol in the other, like Nehemiah, I prepared to ascend the steps of the pulpit" (p. 122). But David Crawford, a friend, leaped from a front pew and wrestled him away from the pulpit stairs. As the two panted and struggled there behind the pulpit, Crawford told Boucher that he knew of at least 20 men in the church under orders to shoot him if he tried to preach. Two other friends joined Crawford to prevent the preaching. The congregation was by this time in an uproar. There were those who shouted that Boucher should be expelled; others that he should be allowed to deliver his sermon. Osborn Sprigg, the leader of the armed men, steered a group to the front of the church, and they surrounded Boucher and his defenders.

As Boucher tells it:

> There was but one way to save my life. This was by seizing Sprigg, as I immediately did, by the collar, and with my cocked pistol in the other hand, assuring him that if any violence was offered to me I would instantly blow his brains out, as I most certainly would have done. I then told him that if he pleased he might conduct me to my horse, and I would leave them. This he did, and we marched together upwards of a hundred yards, I with one hand fastened in his collar, and a pistol in the other, guarded by his whole company, whom he had the meanness to order to play on their drums the Rogues' March all the way we went, which they did. (p. 123)

Because of Boucher's importance during the American Revolution, I shall in this essay trace his career, make note of his writings, and pay particular attention to the most important of his sermons, that delivered on July 24, 1775. I shall describe that piece of rhetorical discourse, comment on some of Boucher's strategies in devising it, and speculate briefly concerning its philosophical underpinnings.

Now, who was this Jonathan Boucher? He was one of approximately 80,000 loyalists or Tories who departed from America as a result of the revolution in contrast to some 400,000 who adjusted to the circumstances and stayed on.[2] He has been called one of the foremost preachers of his time, yet strangely enough his name does not appear in Robert T. Oliver's (1965) *History of Public Speaking in America* nor in De Witte T. Holland's (1969) *Preaching in American History.*

Jonathan Boucher was born in Blencogo in the parish of Bromfield, Cumberland, in England, on March 12, 1738. He was the son of a local statesman who kept an ale house and ran a local school. Despite the fact that the family was pitiably poor, Boucher was able to attend Wigton grammar school. Subsequently he was employed as an usher[3] in a private school in the town of St. Bees while continuing his own studies under the direction of its headmaster, the Reverend John James. In 1759, at the age of 21, Boucher emigrated to America, where he became the private tutor to the sons of John Younger, a Virginia merchant. Two years later, in 1761, he returned to England to take holy orders, and in 1762 he was ordained by the Bishop of London. He returned to America and was placed in charge of Hanover parish in King George's County, Virginia, and later of St. Mary's parish in Caroline county in that same state.

In 1770, he became successively rector of St. Anne's in Annapolis and Queen Anne's in St. George's county, Maryland. He continued teaching and tutoring. One of his students was Jackie Custis, George Washington's stepson. According to an account in *The Dictionary of National Biography* (1968), "he [Boucher] lived in intimate friendship with Washington. They often dined together and spent many hours in talk; but the time soon came when they stood apart."[4] Nonetheless, the volume of sermons Boucher later published was dedicated to our first President.

In recognition of his forceful advocacy of the establishment of an American episcopacy he was awarded an M.A. degree by King's College, now Columbia University. A further honor accorded him was his appointment as chaplain of the Maryland House of Delegates which led to his acquaintance or friendship with many leaders of both Maryland and Virginia, including Governor Eden.

Sometime in the autumn of 1775, at the age of 37, Boucher was forced to return to England. There the government awarded him a pension for struggling in opposition to the revolution. Ten years later, in 1785, he was appointed vicar of Epsom, Surrey. There he spent the last years of his life compiling a glossary of provincial and obsolete words (1833), which Professor Allan Walker Read (1933) of Columbia University regards as important in that it reveals the ways in which many American Indian and Yankee terms entered the English language.[5] This work of Boucher's was purchased from his family in 1831 by the proprietors of the British edition of Webster's *Dictionary* with the intent of making it an appendix.

Boucher died on April 27, 1804. Besides the collection of 13 sermons dedicated to Washington, published in 1797 under the title *A View of the Causes and Consequences of the American Revolution between the years 1763 and 1775*, and the glossary just mentioned, Boucher contributed to William Hutchinson's two volume *History of the County of Cumberland* dated 1794. His letters to George Washington, edited by W. C. Ford, were brought out in 1899. The list of his major writings is completed by the addition of his reminiscences which cover the years 1738–1789. They were edited by his grandson, Jonathan Bouchier, and were published in 1925 and again in 1967.

To complete the portrait of Boucher, we may look to the impressions of him recorded by one of his wives, which are included in the reminiscences:

> In person, inelegant and clumsy, yet not rough and disgusting; of a dark complexion, and with large but not forbidding features. Of a thoughtful yet cheerful aspect; with a penetrating eye, and a turn of countenance that invites confidence and begets affection. Manners—often awkward, yet always interesting; perfectly untaught and unformed, conformable to no rules, yet never unpolite; incapable of making a bow like a gentleman, yet far from incapable of thinking, speaking or acting in a manner unbecoming a gentleman. Never knew a person of so low an origin and breeding with so high and improved a mind; a thorough gentleman as to internals and essentials, tho' often lamentably deficient in outward forms. (Boucher, 1925, p. 80)

And what was it that Boucher accomplished that Sunday in July in his sermon, numbered 13 in the collection? Let me sketch out the content of this very long sermon and indicate some of the tactics Boucher employed in it.[6]

His text was Nehemiah, Chapter 6, verses 10 and 11:

> Afterward I came unto the house of Shemaiah, the son of Delaiah, the son of Mehetabeel, who was shut up: and he said, Let us meet together in the house of God, within the temple, and let us shut the doors of the temple; for they will come to slay thee, yea in the night they will come to slay thee. And I said, should such a man as I flee? and who is there that, being as I am, would go into the temple to save his life? I will not go in.[7]

Boucher begins by saying that the book of Nehemiah might be called the Patriot's Pattern. In the first several of the 38 pages of the sermon he talks of his treatment on Thursday and of the insults and indignities he has suffered. He promises to touch upon them again before concluding. He then states his thesis:

> After this particular application to those by whom the wrong is done, my aim was to suggest to those of more quiet spirits, who are the objects of

these wrongs, some suitable words of comfort, such as may support them
under their impending trial. (Boucher, 1797, p. 563)

Before addressing himself to the rebels and to the loyalists, he says
it is proper to take a more immediate view of Nehemiah. Then, for some
16 or so pages, about half of his sermon, he tells the story of Nehemiah's
trials and tribulations—the conspiracies against him and of his betrayal
by Shemaiah. He describes his steadfast faith in God through these trying
times and Nehemiah's refusal to take refuge in the temple.

Now, clearly what Boucher intended was an analogy. He implied a
comparison between himself and Nehemiah. Going into the temple meant
joining ranks with the revolutionaries or putting oneself in the position of
seeming to join them and by doing so probably to be slain. In the telling
of the story, Boucher could say some things indirectly which he could not
say outright. He injects such barbs as: "The children of misrule will too
probably long continue like the assailants of Nehemiah, to try the fortitude of friends of order" (p. 567), and "The perseverance of bad men
engaged in a bad cause, is almost proverbial" (p. 567). Nehemiah's ene-
mies, he says, were "cruel, because they were cowards; and they were cowards because they were wicked" (p. 568). "Attending more to what seemed
expedient, than to what they should have known as their duty, many of
them swallowed the bait that was so artfully thrown out to lure their
destruction" (pp. 568–569) and "their fears blinded their judgments" (p.
569). But he shows that such fear ends in disaster:

It is a strange feature in the character of weak men, that, having themselves
once been seduced from their duty, it seems to afford them something like
relief to seduce others; as if a disease could be cured by infecting another
with it. (p. 572)

Following the long story of Nehemiah, the second section of the sermon begins at the eighteenth of 38 paragraphs, almost halfway through
the sermon. The prelude to the actual words directed to those in the congregation who would revolt and to those who would not would seem to
indicate that Boucher in this instance may have wished to soften up his
listeners by a long noncontroversial narrative. Even though he did inter-
ject thrusts at his opponents along the way, he may have felt that a long
introductory section was necessary before direct confrontation. We must
note, though, that the introduction attempts to prove the thesis as it goes.

In this second section, Boucher begins by saying:

Addressing myself, then, according to the proposed plan of my Discourse,
now more particularly to those persons in the community, who, either
through passion and prejudice, or through mistaken principles of policy,
pursue with such unrelenting vigor those of their brethren who cannot

adopt or even approve all of their measures, I set out with observing, that for one party to persecute another, merely because of a difference of opinion, is a crime that is much aggravated by the reflection, that there is no temptation to the commission of it, but such as a generous mind must abhor. (p. 577)

He goes on to say that no man was ever made a convert to any opinion by compulsion, and:

There is a principle in our natures which revolts at the idea of being driven. . . . Conviction results only from arguments that will bear to be reflected and deliberated on: whereas to be violent and overbearing only makes men more tenacious to their preconceived opinions; as trees are said to spread their roots, and take faster hold of the ground by being planted in situations where they are much exposed to be shaken by strong blasts of wind. (p. 578)[8]

He then notes that the philosopher Bacon says that good men rarely think exactly alike on any subject and goes on to plead that we all have predilections for deeply rooted doctrines and dogma. Next, he observes that "no two persons can differ more from each other than a man at different periods of his life may differ from himself" (p. 578). In some instances, he says, men's opinions seem to be involuntary and independent even of themselves and we cannot help viewing objects as they are represented to us through the various tempers and capacities anymore than a person placed in a valley can see that which his neighbor on a hill can see.[9]

Boucher then attacks directly telling the audience that "if the allegations of your Committees be well founded, they will assuredly come to nought. They destroy persons who maintain the truth but they cannot destroy the truth itself. They fight against God." Then he quotes Isaiah, Chapter I, verse 2:

Behold, all ye that kindle a fire, that compass yourselves about with sparks! Walk in the light of your fire, and in the sparks that ye have kindled—This shall ye have of mine hand, ye shall lie down in sorrow. (p. 579)

Then: "This, O ye wicked! this is your day. . . . But, remember, that for all this, God shall call you to judgment; and it may yet be your lots also to lie down in sorrow" (p. 580).

After these two long paragraphs, a relatively short passage in contrast to other portions of the sermon, he says a more painful task now takes his attention:

The second object of my Discourse was, to recommend to those of you who like myself, may be so unfortunate as to incur the displeasure of the Committee, fortitude, patience and perseverance in times of trouble. (p. 580)

Dark clouds are gathering over our heads. "But, first, let me warn you not to entertain either a wish, or an hope, that you may be permitted to remain in a state of neutrality" (pp. 580–581). He then gives the example of Titus Pomponius Atticus, who tried to remain neutral in the midst of contending parties only to find that not to be for something is to be against it. In other words, *not choosing* is to choose. There is no such thing as neutrality in time of war.

Following this passage concerning neutrality, Boucher in a strong section throws out motive appeals to fear and then to courage. In juxtaposition he treats of these two motives. He first foresees, "such days of evil awaiting us as may well make men's hearts fail them for fear" (p. 581). He warns men against being confident in their firmness until they are tried and points out that it is no easy trial for a man who is at ease in his possessions to be driven from them. He says that one may be called to part with many comforts and may have to suffer cruel mockings, scourgings, bonds, and imprisonment. He enjoins, "pray, therefore, continually that your patience and your faith may endure in all your persecutions and tribulations" (p. 582). He goes on to talk of fortitude, of the strength that comes from faith. "If any man suffer as a Christian, let him not be ashamed; but and if he suffer for righteousness sake, happy is he" (p. 585). The part one has to act may be difficult but it is not unimportant. He admonishes, "Let your light so shine before men that you may become a light to lighten the wavering" (p. 586).

This passage, directed to the loyalists, is some six paragraphs long and ends with further warning that the way will not be easy:

> For your Committees, Conventions and Congresses, backed as they are by regiments, battalions and armies, are not likely to stop short, til they have overturned government and destroyed and disgraced every man whose principles lead him to wish to preserve it. (p. 587)

In the final 9 pages of the 38, Boucher, at last, talks about himself and his own plight. He tells of surmises that have been whispered into his ear by those opposing him. He says he has been told that unless

> I will forbear to pray for the King, you are to hear me neither pray or preach any longer. [but] ... Entertaining all due respect for my ordination vows ... I will continue to pray for the King and all that are in authority under him; and I will do so not only because I am so commanded, but as the Apostle adds we may continue to lead quiet and peaceable lives in all godliness and honesty. (p. 588)

Boucher notes that if this is to be his valedictory sermon his words will not come from "feigned lips" and points out that the last words are generally regarded as words of importance.

He recounts his coming to the parish, saying that from the first the doors were shut against him.

> My preferment among you, instead of being productive of permanent happiness, as I had fondly hoped it would be, has become one of the heaviest calamities that ever befel [sic] me, even my enemies must be forced to allow that my faults cannot well have been greater than my sufferings have also been. (p. 589)

He becomes here the martyr, admitting in the next paragraphs that he had favored repeal of the Stamp Act—had "joined a giddy and numerous multitude" but "with sincerity in my heart and my Bible in my hand, I sat down to explore the truth" (p. 591). He studied the ancients on government, those who got their materials only from the purest sources: the law of God and the law of the land. The result of the course of readings: the sermons he has preached. He entered into the study to instruct himself and then to instruct them.

> That all my conclusions are certainly true, it would be presumption in me to assert: but you should do me the justice to believe that I think they are true. (p. 593)

Boucher proceeds by talking of his attachments to America and of the ridicule he has suffered because of his English birth. Then finally comes one of the most significant of all passages insofar as the revelation of his basic philosophy is concerned:

> From scraps of conversation, ill understood, and worse related; from mutilated passages of sermons, first heard with prejudice and then commented on by ignorance, positive proof is said to have been obtained that I have preached up the doctrine of *unlimited obedience*. Could this charge really be proved, I should deserve to be proscribed the pale of common sense. It is surprising that men, who pretend to some accuracy both in speaking and thinking should thus confound things and words so totally different as *unlimited obedience* and *passive obedience*. (p. 593)

The remainder of the discourse is like a gallows speech. Indeed, he quotes from that of Lord Verulam which includes the words "I have prayed with my Saviour, that this cup might pass from me!" (p. 595). Boucher's final words:

> And now, thanking you, as from my heart I do, for the respectful attention with which so many of you have long listened to me; and with the warmest cordiality, wishing you patience under your sufferings, and a happy issue out of all your afflictions, I take my leave of you for a season. Brethren, farewell! Be perfect; be of good comfort; be of one mind; live in peace; and the God of love and peace shall be with you. (p. 596)

Let us here turn to the philosophical ideas which served as bases for Boucher's farewell utterance.

In the view of Vernon Louis Parrington (1930), writing in his *Main Currents in American Thought:*

> The single and sacred duty of the subject, Jonathan Boucher was convinced, is faithful obedience to the powers that are set over him. Those powers derive from God and are instituted for the subject's good. It follows, therefore, that the unpardonable sin is rebellion against lawfully constituted authority. "The doctrine of *obedience for conscience* sake," he [Boucher] asserted, "is ... the great cornerstone of all good government." With Daniel Leonarde he makes much of it, but he appeals rather to the sanctions of religion than to the law.[10]

To support his point further, Parrington quotes Boucher from sermons other than the one we have considered:

> Obedience to Government is every man's duty, because it is every man's interest; but it is particularly incumbent on Christians, because ... it is enjoined by the positive commands of God. ... If the form of government under which the good providence of God has been pleased to place us be mild and free, it is our duty to enjoy it with gratitude and with thankfulness. ... If it be less indulgent and less liberal than in reason it ought to be, still it is our duty not to disturb the community, by becoming refractory and rebellious subjects, and *resisting the ordinances of God.* (Parrington, 1930, p. 216)[11]

And in another place:

> Those great and good men, who, *like wise masterbuilders,* have from time to time so *fitly framed together* our glorious Constitution, well knew that *other sure foundation no man could lay* than ... obedience, not only *for wrath,* but *for conscience sake.* (p. 216)[12]

Because this spirit of obedience was openly flouted in America, where every influence made for rough individual liberty, Jonathan Boucher feared for the future.

Boucher said in still other sermons:

> Instituted by God and functioning under divine sanction, government becomes, therefore, a divine instrument, for the security of which He is greatly concerned: Everything our blessed Lord either said or did, tended to discourage the disturbing of a settled government. (Parrington, 1930, p. 216)[13]

> Unless we are good subjects, we cannot be good Christians. Jesus thought it would be better, both for Judea in particular and for the world in general, that ... the people should not be distracted by a revolution, and ... that there should be no precedent to which revolutionists might appeal. (p. 218)[14]

> The very intolerable grievance in government is, when men allow them-
> selves to disturb and destroy the peace of the world, by vain attempts to
> render that perfect, which the laws of our nature have ordained to be imper-
> fect. (p. 218)[15]

Parrington in his final estimate finds Boucher to be the high Tory of the Tory cause in this country. He said that he would not knuckle under to newfangled notions but stood stoutly for God and the King. In sum:

> In laying bare the heart of Toryism, he unwittingly gave aid and comfort to
> the detested cause of liberalism. It is reasonable to assume that such mili-
> tant loyalty to the outworn doctrine of passive submission was a real dis-
> service to the ministry, for it revealed the prerogative in a light peculiarly
> offensive to American prejudices. What a godsend to the liberals was such
> doctrine on the lips of so eminent a divine. (p. 218)

Robert McCluer Calhoon (1973, pp. 218–233) in line with Parrington classifies Boucher as belonging to the group of pre-Revolutionary Loyal-ists which proceeds from the doctrine of authority. The sermon just scru-tinized reflects the beliefs in this doctrine: that authority, that of the king in this instance, is divinely delegated and is in line with the Anglican teachings of the time. The plea is for order which rejects rebelliousness and the use of force and confirms the omnipotence of God. Boucher's proofs in support of this theme are largely Biblical, although at some points he employs examples such as that of Antiochus and the Maccabees and that of Titus Pomponius Atticus. Throughout the sermon Boucher's classical training is evident in the support of his ideas. It is footnoted in Latin at several points, and many of his allusions are to Latin and Greek sources.

Boucher's thinking as reflected by the ideas in the sermon examined derives in part from his reading of Dr. Samuel Clarke, whose rational Anglican philosophy shook his belief in the trinity. Clarke and Boucher as well, as we noted earlier, believed that religious conviction can never be instilled by coercion, fear or manipulation. Yet despite Boucher's belief, it is interesting to note the fear appeals in the latter part of the sermon, where he tells his fellow loyalists that suffering, scourging, per-secution and, perhaps, execution await them.

Aside from Clarke, other influences upon Boucher's thinking were the Reverend Samuel Johnson, George Berkeley, and Robert Filmer. The Anglican clergy, during the first half of the eighteenth century, increas-ingly thought it their duty to eradicate irrational religion and social lib-ertinism through appeals to men's reason in a calm and patient manner and through appeals to their instinctive fondness of order. Johnson made it his life's work to construct such a philosophy. His writings, based largely on George Berkeley's idealism, were a major American effort to explain the nature of knowledge and man in relationship to the world of ideas in order to bolster a theology of rational religion. Johnson fully supported Berkeley's contention that physical matter did not exist, that instead, only ideas made up reality. Certainly, such an approach is reminiscent of Plato. Both Johnson and Berkeley rejected Locke's contention that ideas could enter the mind only through the senses. They held that sense impressions were just another set of abstract ideas and that men perceived truth only in the form of spiritual intuitions.

Robert Filmer (158?–1653) had written the most elaborate rationale for this position by arguing that all creation was patriarchal: God the father of mankind prescribed the authority of kings to rule the people at his behest, and kings upheld the authority of fathers to rule their children. According to Calhoon (1973, p. 229):

> Drawing judiciously on Filmer and other authorities Boucher constructed a comprehensive justification for political subservience. No authority, he argued, was so understanding and properly restrained as that exercised by a loving father over obedient children or a pious king over loyal subjects.

Such was Boucher's rationale for taking what most present-day Americans regard as the unpopular side at the time of the birth of our nation.

NOTES

1. Professor Bronstein's interest in preaching prompts the contribution of this essay to this volume.
2. The estimate is that of Calhoon (1973, p. ix).
3. An usher was an assistant teacher, a sort of tutor and minor functionary.
4. For Boucher's (1925) own account of his relationship with Washington, see *Reminiscences*, pp. 48–50.
5. Professor Read also cited Boucher's contribution to the area of linguistics in a lecture, "Milestones in the Branching of British and American English," delivered at Herbert H. Lehman College of the City University of New York on December 3, 1973. See also Zimmer (1978, pp. 313–326).
6. Anne Y. Zimmer, in her exhaustive and excellent study, *Jonathan Boucher, loyalist in exile*, considers the textual authenticity of the sermon examined here. She finally concludes that "in the absence of any evidence to the contrary, the sermons must be presumed to be Boucher's" (p. 370, footnote). She seems to doubt that the sermon was ever delivered and thinks that all the sermon texts in *A View . . .* were reconstructions (p. 338). She further says: "More importantly, the circumstances under which Boucher left Maryland were not conducive to delivering a farewell sermon. The decision to leave and the plans to implement it were quietly managed. No man of even ordinary

prudence would have delivered a farewell sermon, and Boucher had uncommon good sense. All of the evidence suggests it is wise to treat Boucher's book of sermons as part of his English experience" (p. 341). Her speculations are not of consequence here. Whether Boucher actually delivered the sermon does not negate an analysis of the rhetoric he exhibited. The record as it stands is at least one of his *intended* rhetoric. We may proceed with the assumption that he did say or would have said on the occasion in question what he afterward printed as his "Farewell Sermon."

7. Boucher uses the King James version of the Old Testament. The passage appears in Boucher (1797).

8. Calhoon (1973, p. 222) says that Boucher never sacrificed Samuel Clarke's prominent conviction that religious conviction can never be instilled by coercion, fear, or manipulation.

9. These ideas concerning the perception of objects seem to reflect the influence of George Berkeley noted later in this essay.

10. From Boucher (1797), p. 309.

11. From Boucher (1797), pp. 507–508.

12. From Boucher (1797), p. 306.

13. From Boucher (1797), p. 535.

14. From Boucher (1797), p. 538.

15. From Boucher (1797), p. 543.

REFERENCES

Boucher, J. *Boucher's glossary of archaic and provincial words. A supplement to the dictionaries of the English language, particularly those of Dr. Johnson and Dr. Webster.* London: Black, Young and Young, 1833.

Boucher, J. *Reminiscences of an American loyalist, 1738–1789.* Ed. by Jonathan Bouchier. Boston: Houghton Mifflin, 1925. Facsimile reprint: Port Washington, N.Y.: Kennikat Press, 1967.

Boucher, J. *A view of the causes and consequences of the American Revolution in thirteen discourses preached in North America between the years of 1763 and 1775, with an historical preface.* New York: Russell and Russell, 1797 (reprinted facsimile edition, 1967).

Bronstein, A. J. In defense of sermons. From *In honor of Walter H. Plaut, Rabbi (1919–1964).* Great Neck, N.Y.: Temple Emanuel (used by permission), 1964.

Calhoon, R. M. *The loyalists in revolutionary America, 1760–1781.* New York: Harcourt Brace Jovanovich, 1973. *Dictionary of national biography.* London: Oxford University Press, 1968, p. 911.

Ford, W. C. (Ed.). *Letters of Jonathan Boucher to George Washington, with other letters to Washington and letters of Washington to Boucher.* Brooklyn, N.Y.: Historical Printing Club, D. Clapp and Son, 1899.

Holland, D. T. *Preaching in American history.* Nashville, Tenn.: Abingdon, 1969.

Hutchinson, W. *1793–1797. History of the county of Cumberland, and some places adjacent, from the earliest accounts to the present time: Comprehending the local history of the county, its antiquities, the origin, genealogy, and present state of the principal families with biographical notes, its mines, minerals and plants, with other curiosi-*

ties, either of nature or of art (2 vols.). Carlisle, England: F. Jollie, 1793–1797. Reprinted in 2 vols. as *History of the county of Cumberland* by E. P. Publishing, Ltd., in collaboration with the Cumbria County Library, 1974.

Oliver, R. T. *History of public speaking in America.* Boston: Allyn and Bacon, 1965.

Parrington, V. L. *Main currents in American thought.* New York: Harcourt Brace, 1930.

Read, A. W. Boucher's linguistic pastoral of colonial Maryland. *Dialect notes*, 1933, *6*, 353–360.

Zimmer, A. Y. *Jonathan Boucher, loyalist in exile.* Detroit: Wayne State University Press, 1978.

Index